Transforming Water Management in South Africa

GLOBAL ISSUES IN WATER POLICY

VOLUME 2

Series Editors

Ariel Dinar
José Albiac
Eric D. Mungatana
Víctor Pochat
Rathinasamy Maria Saleth

For other titles published in this series, go to
www.springer.com/series/8877

Barbara Schreiner • Rashid Hassan
Editors

Transforming Water Management in South Africa

Designing and Implementing a New Policy Framework

Springer

Editors
Barbara Schreiner
652 Chamberlain Street
Rietfontein
Pretoria 0084
South Africa
barbara@pegasys.co.za

Rashid Hassan
Centre for Environmental Economics
and Policy Analysis in Africa (CEEPA)
University of Pretoria
Room 2-6, Agricultural Annex
Hatfield
Pretoria 0002
South Africa
rashid.hassan@up.ac.za

ISBN 978-90-481-9366-0 e-ISBN 978-90-481-9367-7
DOI 10.1007/978-90-481-9367-7
Springer Dordrecht Heidelberg London New York

Printed on acid-free paper

Springer is part of Springer Science+Business Media (www.springer.com)

Foreword

The South African contributions to international policy, legislation and the practice of water resources and water services as well as the judicious learning of South Africans from international experience are masterly knitted together in this book by some of the key role players, with a frank and critical analysis of achievements and difficulties before and after liberation in 1994 – against the harsh backdrop of the legacy of centuries of colonialism and decades of apartheid which have shaped the political economy of water in the 'rainbow nation.' This introductory chapter discusses some of the key South African contributions that, from the perspective of the writer, are especially relevant to other developing countries, as well as some international experiences that have been useful to South Africans. The conclusion of the chapter is that South Africa is now well positioned in the international arena to offer a new and important contribution regarding how to deal with the implementation challenges that are affecting most developing countries.

The storyline of the 14 chapters is the contrast of excellent legislation and policies vis à vis implementation impediments. The stage was set by the then Minister of Water Affairs and Forestry, Kader Asmal, when he opened the 1997 FAO Conference to discuss the second draft of the National Water Bill with specialists from Australia, Chile, Mexico, Spain and the USA. He greeted the visitors and stated that he did not want a perfect law… 'only one that works.'

Thirteen years later, the last paragraph of this book reads: 'Such focus on priority areas, and on implementation rather than policy and strategy, will enable the South African water sector to deliver the promise of its remarkable water policy and legislation to the people on the ground. Such delivery will be the true measure of the success of the policy reforms.' The difficulties that South African on-the-ground water management encounters are faced everyday in developing countries where specialists often wait for more information to be gathered and better models to be developed before risking proposing concrete water resource management and protection measures, and non-traditional, but perhaps better, water and sanitation approaches. Paralysis also cripples managers who frequently prefer improving organizational arrangements and even institutional setups and legislation instead of giving implementation a chance.

Perhaps the most useful lesson to be drawn from South African experience is how to deal with the political economy of water. Other countries in the developing world have not experienced an institutionalised racist system as cruel as apartheid, but they certainly have been exposed to colonialism and segregation and still suffer the legacies of that common past. In this respect, the South African experience in dealing with integrated water resources management (IWRM) within such a complex socioeconomic environment is relevant to many developing countries.

One of the main assets of the National Water Act (NWA) of 1998 is that not all provisions of the Act would come into force from the day of enactment, but that it would be implemented in a phased and progressive manner, in separate components over time according to geographical need and as soon as was deemed reasonable and practical. This decision was partly inspired by international experience, shared in the 1997 FAO Conference, such as the experience of the Mexican water law which came into effect the day immediately after its publication on 1st December 1992 and provided only for 1 year, and through regulations for an additional 2 years, to formalise all existing water uses. The formalisation process took, in fact, 10 years. This cross feedback shows how south-to south learning – and not only of success stories but of challenges and unsolved problems – can be far more productive than blindly following 'international best practice' from the so-called developed world. However, not setting deadlines in an environment that is constantly looking for technical and institutional perfection may also be dangerous. Therefore, it would be desirable now to set up implementable work programs with rigorous deadlines for actions as important as establishing Catchment Management Agencies where they are needed.

It is noteworthy how developing countries can influence international IWRM paradigms. For instance, thanks to South Africa's vision and persistence, they now include the social and ecological values of water on an equal footing with its economic value, and embrace concepts such as the primary role of national government, public ownership of water and the need for gradual implementation. In practice, South Africa has demonstrated the benefits of closely linking water resources and water services management, as well as the need to link the water sector with its multi-layered environment, through coordinated government and modern approaches to equitably shared international water. This can be illustrated with a modified 'Global Water Partnership Comb' showing that water development and management do not exist in isolation.

The perception of the authors of most chapters of the book is that the key factor for underperformance both in sustainable water resources management and effective and sustainable water services is the country-wide lack of scientifically and technically well trained and socially sensitive personnel combined with severe financial challenges at the municipal level in particular. The actions required to remove these barriers lie often out of the realm of South African water resources and water services managers. Nevertheless they seem to be in a position to voice their concerns in national debates and help to empower society by

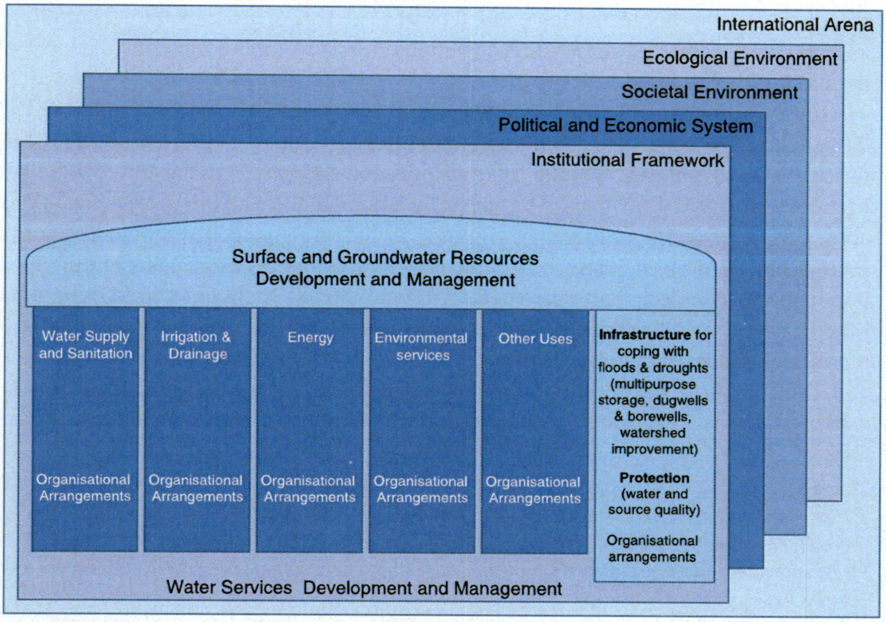

1. Being frank about the consequences of 'business as usual.'
2. Acknowledging capacity limitations for policy implementation.
3. Providing transparent information to ensure integrity and accountability, and counteract vested interests.
4. Challenging unfit macro-policies.

This would trigger a dialogue with the upper level tiers of government responsible for providing the requested additional resources and removing obsolete bureaucratic restrictions to hiring adequate personnel with competitive salaries and opportunities for career development. Of special interest is the educational system that should provide better graduates with sound scientific and technical basic knowledge but also with water resource and water services management skills.

The progress reported in various chapters of this book regarding water resources and water services development and management clearly shows there is the will in different spheres of government to contribute to this process. The dialogue could be structured around various water resources and water services scenarios, showing the need for water reallocation, and/or demand management measures such as wastewater reuse. Each scenario would also make explicit the required financial resources and cooperative government measures. The next step would be to compare these requirements with government's budgets and cooperative programs for similar items during the past 10 years or so. Most probably the trend would be less than required and a request to Parliament for increasing budgets would have to be made. In case this was not feasible under the current level of national income, DWA could make a point for increasing taxes in order to be able to abide by the constitutional and NWA mandates of redressing past inequities, by making the consequences of not having the required moneys clear.

However, increasing taxes is always politically difficult and the goodwill of those who have benefitted from apartheid and colonialism to share with those historically disadvantaged would be indispensable. South Africa, led by Nelson Mandela, in spite of, or thanks in part to, years of blood-shedding, was capable of achieving a peaceful political revolution. Achieving the still-required economic revolution for more inclusive economic development will not be easy, but the rainbow country is closer than other developing countries, such as Mexico, where the inheritance of colonialism and current status is such that we still are in need of both a political and an economic revolution.

Going back to the modified 'GWP comb,' it is useful to bring in a broader aspect of the international backdrop. In his inaugural speech, President Barack Obama said:

> ...The question we ask today is not whether our government is too big or too small, but whether it works –
> ...Our economy is badly weakened, a consequence of greed and irresponsibility on the part of some, but also our collective failure to make hard choices and prepare the nation for a new age.
> ...Nor is the question before us whether the market is a force for good or ill. Its power to generate wealth and expand freedom is unmatched, but this crisis has reminded us that without a watchful eye, the market can spin out of control – and that a nation cannot prosper long when it favors only the prosperous.

These quotes, coming from the first black President of a country which for decades favored the Washington Consensus that nurtured the widespread fantasy that an unregulated private sector and 'slim' government would solve all problems, *inter alia* water problems, are quite relevant since this book clearly highlights lack of government capacity at national, municipal and local levels as one of the most important impediments to sustainable water resources and water services development and management.

Barack Obama after being recently awarded the Peace Nobel Prize must rise to the high expectation he has created with his excellent reconciliatory speeches by taking action to really help bringing peace to the world. Similarly, South Africa must now rise to the high expectation she has generated by successfully dealing with her political economy of water by finding ways to simplify procedures and increase financial resources for ensuring implementation of sustainable water resources and water services development and management.

Héctor Garduño
International Water Resources Planning and
Management Consultant, Mexico City
hector.garduno@live.com.mx

Acknowledgements

Our thanks go to Ariel Dinar who originally came up with the idea for this book, who set us on the path of getting the book together, and who provided guidance and support along the way. Our thanks also go to Jenny Fidler and Phillipa Kanyoka for providing invaluable support in checking the manuscript, dotting the i's and crossing the t's. Thanks to Centre for Environmental Economics and Policy in Africa (CEEPA) for their financial support. Finally, our thanks to our families for their patience with the late nights and the long days that went into pulling this manuscript together.

Acknowledgements

Contents

Biographies of First Authors

Barbara Schreiner is Practice Director for Water Strategy at Pegasys Strategy and Development in Pretoria. She is the former Deputy Director General: Policy and Regulation at the Department of Water Affairs and Forestry. She joined the Department in 1995 as Special Advisor to Minister Kader Asmal. She was involved in the writing of the White Paper on a National Water Policy for South Africa and the drafting of the National Water Act. She holds a Masters Degree in Environmental Science from the University of Cape Town. She has been active in the water sector internationally including as a member of the Global Water Partnership Steering Committee, and the Gender and Water Alliance. She is a member of the Board of the Challenge Programme on Water for Food. She has a particularly interest in gender and poverty aspects of water management.

Barbara van Koppen (Ph.D.) is principal researcher Poverty, Gender, and Water in the Southern Africa Regional Program of the International Water Management Institute, based in Pretoria, South Africa. Her research focuses on pro-poor and gender equitable rural water development for multiple uses and plural institutional and legal frameworks in Africa and South Asia. She has authored and edited four books and over 60 international publications. Before joining the International Water Management Institute, she was a lecturer in Gender and Irrigation at Wageningen University and Research Center, Netherlands. She also worked as a Technical Assistant for the Netherlands Development Cooperation in Burkina Faso.

Chris Moseki works at the Water Research Commission in Pretoria. He has more than 7 years experience as teacher and lecturer. He holds a Bachelor degree in Geology from the University of California Los Angeles and a Masters degree in Geohydrology from the University of the Free State. He has 5 years experience in groundwater development and management as well as another 5 years in water resource management in the Department of Water Affairs in South Africa. He is a member of the International Association of Hydrogeologists, Groundwater Division of South Africa, and a registered earth scientist with South African Council for Natural Scientific Professions.

Eiman Karar is a scientist with more than 16 years experience in the water sector ranging from hard modeling and geomorphological sciences to institutional matters that focus on collective action, organizational transformation, governance, performance bench marking, devolution and empowerment, decentralization and democratic management of water resources. She is a research manager at the Water Research Commission in Pretoria

Gavin Quibell has 21 years experience in the water and environment sector, including 13 years in the Department of Water Affairs in South Africa. He has worked on a wide range of water resources management issues, including water quality management, IWRM and studies on water for the environment. He is currently involved in development projects in the water sector, particularly in Africa. He has worked on a number of donor-funded projects in the water quality and water resource management sectors, and has worked as a freelance consultant to the DWA project on water allocation reform.

Harrison Pienaar (MSc Hydrogeology) started his career in the energy sector with Eskom where he successfully planned and managed a wide range of environmental projects for sustainable linear development. He then pursued his career as a consultant, undertaking and managing numerous large scale projects in integrated environmental management in association with various consulting firms. He joined the Department of Water Affairs in 2001 and currently holds the position of Chief Director responsible for the development of policies, strategies, systems, methodologies and guidelines for resource directed measures (RDM), particularly the Reserve determination, water resource classification and specification of associated resource quality objectives.

Héctor Garduño worked for 30 years (1967–1997) as researcher, private consultant and official of the Mexican Government. After completing his last government task designing and implementing the national water rights administration system, he started his international consulting career by participating in the 1997 FAO Conference in Pretoria to discuss the second draft of the South African National Water Bill. He came back in September 1997 and February 1998 to work with DWA in anticipating implementation issues when they were designing the water rights system and to propose a training program. From July 2004 to July 2005 he returned to South Africa where the World Bank Institute invited him to write a case study on South Africa's experience in implementing Integrated Water Resources Management within its new water policy (Garduño and Hinsch 2005[1]). On July 2006, he came back as member of the International Review Panel for the Institutional Review of the Water Research Commission and on November 2008 he participated in the Round Table Discussion on 'Water for Development' invited by DWA. Since 2000 he has been a core member of the World Bank's Groundwater Management Advisory Team (GW-MATE) working in Asia, Africa, Latin America and the Middle East in Groundwater Resource Planning, Management and Institutions.

[1]Garduño, H and Hinsch, M. 2005. IWRM Implementation in South Africa: Redressing past inequities and sustaining development with a view to the future. World Bank Institute. Washington D.C.

Jayant Bhagwan is Director of the Key Strategic Area of Water Use and Waste Management at the Water Research Commission in Pretoria. He is a civil engineer with post-graduate degrees in Municipal Engineering and Tropical Public Health Engineering. He started his career with a small NGO providing water supply and sanitation support to disadvantaged and poor communities. He then moved to the CSIR focusing on research and development on water and sanitation issues. He has contributed to shaping the research portfolio on the institutional and management aspects of water services delivery, has contributed to many initiatives in the preparation of the new water legislation and serves on a number of key committees. He was Chairperson of the Minister of Water Affairs and Forestry National Water Advisory Committee and a Steering Committee Member of the Water Supply and Sanitation Collaborative Council.

Johan van Rooyen is the Director: National Water Resource Planning in the Department of Water Affairs responsible for strategic level planning at the national and catchment scale. He is a civil engineer by training and has been working in the Department of Water Affairs since graduation, a total of 34 years of which the last 27 years were in planning. His experience includes the planning of water resource infrastructure from small scale rural projects to mega projects like the Lesotho Highlands Water Project; the development of water resource management strategies at both the catchment and national scale, as well as for shared international basins; and managing the development of water resource models for the effective operation of systems as well as for allocation of water.

Kathy Eales has worked in the water sector since the early 1990s from within DWA, in consulting, in Africa's largest water and sanitation NGO, and in municipal government. Her work spans research, policy and strategy development, project implementation, and now regulation. Kathy has played a prominent role in shaping debate around ways of remedying South Africa's sanitation service backlogs, and has worked on several international research projects. She has worked for the City of Johannesburg, where she was responsible for regulatory oversight of the City's corporatised water services provider, Johannesburg Water, and related policy research, and for the Water Research Commission. She now works independently as researcher and consultant, focused primarily on the institutional dimensions of effective water and sanitation services.

Mahomed Vawda is Director: Water Resources Pricing and Financing at the Department of Water Affairs. He comes from a commercial and financial background with experience in both the public and private sector. His experience includes Accountancy, Finance, State Subsidies, Taxation and International Marketing and Investment. Mr. Vawda joined the Department of Water Affairs in January 2004, where one of his responsibilities is the Raw Water Pricing Strategy.

Mark Dent is Director of the Environmental Management Programme at the Centre for Environment, Agriculture & Development (CEAD) at the University of KwaZulu-Natal (UKZN). A water resources engineer by profession, Mark has enjoyed a varied career as lecturer, researcher, manager, consultant and Director of the Graduate School of Business, UKZN. He holds a Ph.D. in water resources engineering and a Masters in Business Leadership from UNISA. Mark's passion is for systems integration, continuous improvement of teams, leadership development and organisational change and more particularly the role that information and our approach to it, plays in the knowledge management, that is essential to success

Rashid Hassan is Professor of Natural Resource and Environmental Economics and Policy and Director of the Centre for Environmental Economics and Policy in Africa (CEEPA), Faculty of Natural and Agricultural Sciences, University of Pretoria in South Africa. He has taught and supervised academic and non-degree training and research of many students from Africa and other parts of the world in economics and policy of managing natural resources. He has published a number of contributions on water resource management and economic policy in key national and international journals and books printed by various publishers.

Reginald Tekateka is an independent consultant who was previously Specialist Advisor in International Relations at the Department of Water Affairs. Prior to that he was Chief Delegate to the Lesotho Highlands Water Project. During his term of office with LHWP his responsibility shifted to include dealing with international water issues including in SADC, NEPAD, CSD and AMCOW. During period this period he was the RSA representative in the SADC senior officials water sector. He is now a member of the AMCOW TAC focusing particularly in the finance working group within the EU Water Initiative. His background and training is international relations.

Contributors

Saliem Fakir
WWF, Millennia Park, 16 Stellentia Avenue, Stellenbosch,
7630, South Africa
sfakir@wwf.org.za

Marna de Lange
Socio-Technical Interfacing, PO Box 1250, Ladanna,
0704, South Africa
marna@global.co.za

Robyn Stein
Mandela Institute, Department of Law, University of the Witwatersrand,
Private Bag 3, Wits 2050, South Africa
rstein@icon.co.za

Ashwin Seetal
Golder Associates Africa (Pty) Ltd, PO Box 6001, Halfway House,
1685, South Africa
aseetal@golder.co.za

Noxolo Ncapayi
Department of Water Affairs, 183 Schoeman Street,
Pretoria, South Africa
ncapayin@dwa.gov.za

Antonia Belcher
Independent Consultant, P.O. Box 195, Stellenbosch,
7599, South Africa
toni.b@iburst.co.za

Dana F. Grobler
BlueScience Consulting cc, PO Box 54, Private Bag X 8,
Kuilsriver 7579, South Africa
dana@bluescience.co.za

Gugu Mazibuko
Pegasys Strategy and Development, 193 Bronkhorst Street, Brooklyn,
Pretoria, South Africa
gugu@pegasys.co.za

Thomas Gyedu-Ababio
South African National Parks, Kruger National Park,
Mpumalanga, South Africa
ThomasGa@sanparks.org

Derek Weston
Pegasys Strategy and Development, 193 Bronkhorst Street, Brooklyn,
Pretoria, South Africa
derek@pegasys.co.za

Toriso Tlou
Tlou Consulting, PO Box 1309, Pretoria,
0001, South Africa
toriso@tlouconsult.co.za

Cornelius Ruiters
Department of Water Affairs, 183 Schoeman Street,
Pretoria, South Africa
ruiterc@dwa.gov.za

Nicola King
Mintek, Private Bag X3015, Randbung
2125, South Africa
NicolaK@mintek.co.za

Mike Muller
School of Public and Development Management, University of the Witwatersrand,
Private Bag 3 Wits 2050, Johannesburg, South Africa
mikemuller1949@gmail.com: mike.muller@wits.ac.za

Abbreviations

AMCOW	African Ministers' Council on Water
ANBO	African Network of River Basic Organisations
ANC	African National Congress
ARC	Agricultural Research Council
AU	African Union
BBM	Building Block Methodology
BEE	Black Economic Empowerment
BNR	Biological Nutrient Removal
CBA	Cost Benefit Analysis
CEDAW	Convention on the Elimination of all forms of Discrimination Against Women
CMA	Catchment Management Agency
CMF	Catchment Management Forum
CMS	Catchment Management Strategy
CSIR	Council for Scientific and Industrial Research
CSM	Combined Services Model
CUC	Capital Unit Charge
DBSA	Development Bank of Southern Africa
DEAT	Department of Environmental Affairs and Tourism (in 2009 this Department became the Department of Environmental Affairs)
DWAF	Department of Water Affairs and Forestry (in 2009 this became the Department of Water Affairs, DWA, as it had been prior to 1994)
EFR	Estuarine Flow Requirements
EMF	Environmental Management Framework
EMP	Environmental Management Plan
EWR	Environmental Water Requirements
FAO	Food and Agricultural Organisation
FETWater	Framework for Education, Training and Research in Water
GEF	Global Environmental Facility
GTZ	German Technical Assistance
GWP	Global Water Partnership
HDI	Historically Disadvantaged Individual
ICOLD	International Commission on Large Dams

ICT	Information and Communications Technology
IDP	Integrated Development Plan
IFR	Instream Flow Requirements
IMP	Industries, mining and power generation
IRP	Integrated Resource Planning
ISPs	Internal Strategic Perspectives
IWRM	Integrated Water Resource Management
LIMCOM	Limpopo Basin Commission
LWHP	Lesotho Highlands Water Project
MC	Management Class
MUS	Multiple Use Systems
NEPAD	New Africa Partnership for Development
NWA	National Water Act
NWRIA	National Water Resource Infrastructure Agency
NWRI	National Water Resource Infrastructure
NWRS	National Water Resource Strategy
O&M	Operation and Maintenance
OKACOM	Okavango River Basin Commission
ORASECOM	Orange Senqu River Basin Commission
PWC	Permanent Water Commission
R&D	Research and Development
RDM	Resource Directed Measures
RDP	Reconstruction and Development Programme
ResDSS	Reserve Decision Support System
RHP	River Health Programme
RQO	Resource Quality Objectives
RWP	Regional Water Policy
RWQOs	Resource Water Quality Objectives
RWS	Regional Water Strategy
SADC	Southern African Development Community
SALGA	South African Local Government Association
SAREP	South African Rainfall Enhancement Programme
SDC	Source Directed Controls
SDF	Spatial Development Framework
SFR	Streamflow reduction
TCTA	Trans Caledon Tunnel Authority
TDA	Transboundary Diagnostic Analysis
TPTC	Tripartite Permanent Technical Committee
WC/WDM	Water Conservation/Water Demand Management
WDCS	Waste Discharge Charge System
WMA	Water Management Area
WMO	World Meteorological Organisation
WRC	Water Research Commission
WRCS	Water Resource Classification System
WRMC	Water Resource Management Charge

WSA	Water Services Authority
WSDP	Water Services Development Plan
WU&WM	Water Use and Waste Management
WUA	Water User Association
Currency	All figures are presented in US$, calculated at a conversion rate of ZAR7.4: 1US$

List of Boxes

List of Figures

List of Tables

Chapter 1
The Political, Social and Economic Context of Changing Water Policy in South Africa Post-1994

Barbara van Koppen, Barbara Schreiner, and Saliem Fakir

Abstract This chapter describes the political, social and economic context in which South Africa's water reform was designed and implemented. The water reform was part of the nation's wider transformation after 1994 from white minority rule and territorial and institutional segregation, to a democratic, non-racial state. This implied a major challenge to redress the legacy of gross inequities in access to water for domestic and productive uses and the persistently high poverty levels, especially in the rural areas. For a better understanding of the continuities and changes from the past for all aspects of water reform discussed in this volume, the history of water development and management in apartheid South Africa is traced. This encompasses the removal of land and water rights from black South Africans by the early 1900s; the hydraulic mission for white agriculture throughout the twentieth century; and the emergence of the centrally planned, urban-industrialized water economy from the 1970s onwards. Many concepts that would globally be seen as 'best practice' Integrated Water Resource Management according to the Dublin principles of 1992 originate in that era. The chapter concludes by introducing the subsequent chapters in this light.

Keywords History • Integrated water resource management • Water infrastructure • Water law and policy • Water reform • Equity • Bill or rights • Catchment mangement agency • Water user association

---

B. van Koppen (✉)
International Water Management Institute, South Africa
e-mail: b.vankoppen@cgiar.org

B. Schreiner
Pegasys Strategy and Development, South Africa
e-mail: barbara@pegasys.co.za

S. Fakir
Stellenbosch University, South Africa
e-mail: saliem@sun.ac.za

B. Schreiner and R. Hassan (eds.), *Transforming Water Management in South Africa: Designing and Implementing a New Policy Framework*, Global Issues in Water Policy 2, DOI 10.1007/978-90-481-9367-7_1, © Springer Science+Business Media B.V. 2011

1.1 Introduction

The water sector in South Africa has experienced radical changes since liberation in 1994 and the subsequent introduction of new policies and legislation, namely, the White Papers on a National Water Policy for South Africa (RSA 1997) and on Water Services (RSA 1994), the National Water Act (RSA 1998) and the Water Services Act (RSA 1997a). Significant changes in water management were driven by the need to create more socially just, economically efficient and environmentally sound water management and allocation regimes in the country, in addition to bringing safe water and sanitation to the vast number of South Africans left unserved by the apartheid regime. The period of policy and legislative reform was followed by a decade of implementation of new approaches spanning most aspects of water management. This included changes in the approach to water-use authorisation and allocation, introducing pro-poor and environmentally friendly approaches, changes in institutional arrangements, and major shifts in managing demand and charging for water use.

While it will take many years to achieve the full effect of the changes, valuable lessons have already been learned in the process. This volume aims to analyse and document these experiences for the benefit of water managers and policy makers in the international community. The analysis is intended to be critical, referring to both successes and failures in the various areas and, where possible, drawing lessons from the South African experience.

In order to fully appreciate the policy reforms after 1994 in the water sector, it is important to understand how these reforms are closely linked to the wider political economy that drives water development and management, which, in its turn, shapes the country's development path. In South Africa the reforms were embedded in the political and economic changes in the country as a whole, when the African National Congress (ANC) became the first democratically elected government of South Africa. Until 1994, race, gender and class were the dominant organising principles of South Africa's political economy and water management. Racial discrimination penetrated all spheres of life, and advantaged white men over Africans, so-called coloureds, Asians, and white women. In this book 'black' refers to Africans, coloureds and Asians, as defined in the Broad-Based Black Economic Empowerment Act (Act 53 of 2003).

The legacy of this history was that, at the dawn of democracy, an estimated 12 million blacks did not have access to safe water, and around 20 million did not have adequate sanitation – this of a population in the order of 36 million. Moreover, hardly any black men, let alone black women, used any significant quantities of water for productive uses, or had a formal water entitlement in their own name. Water resources were concentrated among the white population. With technical sophistication, whites harvested most of South Africa's limited annual rainfall. In the stressed Olifants Water Management Area, for example, half a percent of the population – virtually exclusively white men – controlled 95% of the water resources (Cullis and Van Koppen 2007). And yet, many of the ideas on improved water resources management that are currently seen as best practice Integrated Water Resources Management had already emerged under the apartheid regime in the 1970s, two decades before the Dublin Principles of 1992.

Profound political changes took place after 1994 in terms of race and gender. However, there was also much continuity, especially in the nature of South Africa's capital-intensive urban industrialized and large-scale agricultural economy. After 1994, this economy stopped serving the white minority only and included an increasingly racially-mixed middle-class and, indirectly, the many poor who moved to the fringes of the urbanized areas. However, inequities in wealth continued to widen, a trend that started in the 1970s. As a result, poverty remains rampant, especially in the former homelands. For deriving generic lessons on the link between water and poverty eradication, the critical question is whether and how water reform followed or reversed these wider economic trends.

The remainder of this first chapter aims to sketch water management pre-1994 and how this was linked to the socio-economic and political context of colonization and apartheid. This allows, firstly, a better understanding of the legacy of inequities that the water reform attempted to redress. Secondly, this analysis provides a basis for better discerning the changes and continuities in the new measures analysed in the remainder of this volume. In the last part of this chapter, the subsequent chapters of this volume are briefly introduced in this light.

1.2 Pre-1970s: Resource Dispossession and the Hydraulic Mission for White Agriculture

1.2.1 Resource Dispossession

The reasons for the fact that, by 1994, almost no black person had water entitlements in their own names, go back to the arrival of European colonisers in 1652 and the subsequent centuries of resource appropriation through warfare and divide-and-rule alliances with mutually competing tribal groupings. As a result, by 1900, the colonial powers had dispossessed Africans of large parts of their land, water and mineral resources. When the patriarchal elite 'alliance of maize (Afrikaner agro-pastoralist landlords) and gold (British mining houses and bankers)' joined hands to form the Union of South Africa in 1910, it consolidated this resource grab and enforced it through 'lawful' institutional arrangements across South Africa (Terreblanche 2002). The Union also denied the vote to black people and white women, except for property-owning blacks in the Cape Province. A white suffrage movement ensured white women's voting rights in 1930 (Saunders and Southey 2001), while Africans obtained the vote only in 1994.

New land and water laws of the white rulers institutionalised this resource appropriation and enforced it through military and police force. Land Acts relegated all Africans, who constituted at least 80% of the population, to so-called Native

Reserves which initially covered only 9% of the total area. This expanded to 13% under the Land Act of 1936 (Lahiff 1999). After the apartheid government came to power in 1948, these Reserves were consolidated into nine so-called 'homelands', through another wave of forced removals. This race-based territorial and institutional segregation eroded African women's indigenous status (already minors), and their status as independent producers, even further. Traditionally, cropping and some craft and small enterprises were women's domain, while men were engaged in livestock husbandry. While land belonged to the men of a clan, after land clearing, which was usually done by men, women farmers had quite strong land use rights, controlled crop cultivation and had a say over the use of produce, which was mainly for family consumption (Monnich 1967; Delius 1983). Later, formalisation of this indigenous land tenure by the colonial government into 'Permissions to Occupy' (PTOs) excluded women.

In the rural parts of the white areas, the Land Acts eroded the limited rights that black tenants used to have under rental arrangements with the whites who had registered this land as white-owned property, and downgraded tenants' labour status to the most exploited wage labour (Bundy 1988). In the urban parts, a suite of legislation, including the Native Urban Areas Act (Act 21 of 1923) and the Native (Urban Areas) Consolidation Act No 25 of 1945 resulted in separate residential areas for Africans, coloureds, people of Asian origin and people of European descent, and the creation of 'dormitory' townships for the black labour force.

The dispossession of water rights was closely related to the territorial segregation. The Irrigation and Conservation of Water Act of 1912 ingrained the 'lawful' appropriation of most of the nation's water resources by whites. After internal debates whether to adopt the British common-law riparian rights systems or the civil-law Dutch-Roman permit systems, riparian rights were chosen (Lewis 1934). Riparian land ownership already accrued exclusively to whites under the British title deed system (Terreblanche 2002), so, by tying water rights to land, property-owning whites appropriated the related water resources with a stroke of the pen. Other Acts, like the Gold and Mineral Acts and the 1956 Water Act, stipulated special powers to control both water and mineral resources within the homelands (DWA 1956).

The erosion of Africans' control over their resource base not only boosted whites' exorbitant resource control, but also served to provide cheap labour for the mines (men only) and farms (women with the lowest wages, and men) and evolving manufacturing and industries (mainly men) and domestic work (mainly women). A battery of increasingly rigorously enforced labour laws ensured that blacks became an ultra-exploitable labour force (Thompson 2001; Terreblanche 2002). The 'black peasant road to capitalism' (Crais 1992, cited in Terreblanche 2002), which had taken off around the turn of the twentieth century was destroyed by the erosion of resource rights and growing competition from heavily-subsidised white farmers. Black women continued subsistence production in the Native Reserves without state support, justifying even lower wages for males, and saving the state expenditure on raising the new generation of labourers, caring for the sick, and hosting pensioners.

Resistance to racial and class discrimination against black South Africans mounted, along with the political developments of white South Africa. The African National Congress was launched in the same year that the Irrigation and Conservation of Water Act was passed, with the express intention of promoting the rights and freedom of the African people.

1.2.2 The Hydraulic Mission for White Agriculture

Throughout the twentieth century, whites took up the rapid development of storage and conveyance infrastructure. The state played a critical role in identifying favourable dam sites and other hydrological and technical opportunities, and in stimulating long-term investments in collective infrastructure that would never have been made without a public agency (O'Meara 1983; Backeberg 1994).

Water development for mining and urbanisation was organised and financed from the internal resources of mining houses or white local government. Self-financing or repayment of capital investment costs was widespread practice among these localised collective water supplies for municipalities, mines, and later for industries and electricity generation. The Union state almost exclusively focused its water support on stimulating white agriculture.

The Irrigation Department rolled out across the Union what had already been experimented on in the nineteenth century: stimulating groups of farmers to form Irrigation Boards as a condition for obtaining loans and technical advice. From the 1930s onwards, the state itself started financing and constructing Government Water Schemes for lease, especially to poor white farmers (Backeberg 1994). These investments quieted the latter's demands for race-based privileges (Turton and Meissner 2003) and consolidated encroachment in remote areas, where white ownership of land was still largely a paper-law affair.

Initially, there was some intention that investments in irrigation and agriculture should be commercial loans that were to be repaid by farmers. However, by the 1920s this was abandoned because there were no takers for such loans, while the national interests of higher productivity and food security, 'as a national assurance against drought', were increasingly well-established. Therefore, grants of one third of capital cost were given with soft loans for the remainder capital costs. From 1946 onwards even private farmers could obtain such grants and loans. Repayment was notoriously weak (Backeberg 1994). Until the 1970s the then Department of Water Affairs had no accountancy service to administer repaid loans (DWA 1970).

Black South Africans did not benefit from similar public investments, neither in water services for domestic uses nor in water development for productive uses. Without any incentive to boost the autonomous productivity of a black labour force, and with minimum incentive to share benefits with them, the hydraulic mission for white agriculture deepened inequities in access to and control over water resources.

1.3 Post-1970s: The Emergence of a White Urban-industrial Water Economy

1.3.1 Growing Inequities

The shift from the extractive white agriculture- and mining-based economy to a white minority economy based on manufacturing, industrialization and services after the Second World War and especially from the 1970s onwards, created large-scale unemployment, a reality grossly ignored by the white water sector, which focused on its own transformations into a white urban-industrial water economy. Before the 1970s, cheap labour had been a relatively scarce resource and remained in demand. However, in the second half of the twentieth century, when the white economy moved from mineral extraction and agriculture to capital-intensive and high-tech industrialisation and modern services provision, it ceased to absorb the under-educated labour force. By 1970, more than 20% of the potential labour force was unemployed. The figures rose to almost 40% in 1995, partially due to population growth in these 25 years, with an increase of 42% in the potential labour force. At the same time, employment grew by only 14%. The number of non-African workers employed *increased* by 45%, while the absolute number of Africans employed *decreased* by 3.8% (Terreblanche 2002).

Growing unemployment was accompanied by increasing differentiation among Africans in which a minority gained while the majority lost even more. With the African trade-union strikes and the abolishment of discriminatory labour legislation in the 1980s, the income of the top 20% of African households increased by 38% from 1975 to 1991. However, in the same period, the income of the other 80% of Africans declined, with the income of the poorest 40% declining by 42% (Terreblanche 2002).

African women tenants and farm labourers were hit hardest by the modernisation of the white economy. The number of farms declined from about 120,000 in 1952 to some 60,000 in 1983 (Fényes and Meyer 2003). The number of male farm workers decreased by 38% from 1970 to 1995, and the number of women by 72% (Vink and Kirsten 2003; SSA 2000, cited in Vink and Kirsten 2003). With the loss of work went evictions from farms, which, with forced removals driven by the apartheid government, accelerated congestion in the already over-crowded homelands.

1.3.2 The Creation of the Urban-Industrial Water Economy: IWRM avant la lettre

Changes in the water sector served the purposes of the changing white economy, entirely ignoring its impacts on the black majority of South Africans. With growing economic diversification, the emerging other water use sectors started to challenge the strong riparian claims of landowning farmers. With the 1956 Water Act, the Department of Irrigation became the Department of Water, and included growing non-agricultural sectors' water needs.

In the 1960s industrialisation and energy provision boomed and tertiary sector services grew, especially in today's Gauteng and the eastern adjacent coal fields with Eskom's coal-fired electricity generation and petro-industries. This supported an agricultural transition in which many white farmers found better employment in the industrialising economy. Only a few farmers stayed and mechanized their expanding farms. General state support for agriculture dwindled until it was further deregulated in the 1980s.

Water was scarce on the elevated 'high-veld' of South Africa's new economic hub, where only two basins provide head waters. In 1970, a 'Commission of Enquiry into Water Matters' was created to assess the new needs. This Commission introduced many new concepts for water management in South Africa (DWA 1970). Indeed, the apartheid state became a proponent of Integrated Water Resource Management '*avant la lettre*'.

Long before the Dublin principles would define hydrological boundaries as the most appropriate boundaries for water management, the Department of Water Affairs realized their importance in order to take water *out of* basins through basin transfers. Interbasin transfers were needed in order to provide water to Gauteng. Some transfer from the Vaal had already existed, governed by the Rand Water Board, shortly after its creation in 1903. After the 1970s transfers expanded drastically from all basins surrounding this economic hub. Plans were initiated to tap water from Lesotho and studies explored bringing water from the Zambezi. Distances for water conveyance increased and only less-favourable dam sites remained for further development (DWA 1970, 1986). A new niche emerged, of centrally co-ordinated management according to basin boundaries, to move water over large distances both within basins and, even more, across basins. This niche was taken up by the Department of Water Affairs as centralized co-ordination was critical for meeting economic needs that crossed hydrological boundaries. However, in the territorially divided country, where relationships with the homelands were governed according to the Helsinki principles for transboundary water management, there was no notion of catchment management agencies as yet,

In the new competition for water, and even more for state funds for water development, the heavy subsidization of white agriculture was challenged for the first time in the water sector's history. However, apart from stopping any new construction of Government Water Schemes, little changed in reality, not even when state budgets further reduced due to the weakening economy and the major expenditure required for oppressing the growing anti-apartheid movement. The white agricultural sector, an important constituency of the apartheid government, could resist change.

In this context, various arguments were coined to discourage state support to agriculture. The 1970 Commission proposed Irrigation Management Transfer, with generous phasing-out subsidies, in order to further reduce state expenditures (and not because of any assumed 'better management' as promoters of irrigation management transfer would later claim in other countries). The desirability of 'economic viability' and repayment of loans or, in any case, operational costs ('the user pays'), as had existed in the early twentieth century, was revived (DWA 1970, 1986). DWA appointed its first accountants to ensure repayments (DWA 1970, 1986). The Commission also took up water demand management and greater water efficiency, again, in particular, for irrigation: 'The Commission is of the opinion that the struggle towards a 25% saving of irrigation water per unit of production can succeed' (DWA 1970). The Commission revived the call for volume-based payment, sliding scales, and intra-scheme water trade, as the 1952 Irrigation Commission had already recommended (DWA 1952, cited in Backeberg 1994). Realising the merits of 'virtual water' well before its global invention, the 1970 Commission proposed to import food and other products rather than extend irrigated areas for national food security (DWA 1970).

The Commission also argued in favor of water supplies to the urban-industrial economy by promoting the notion of water as an economic good by underlining that water in a "free" economy has to be developed, allocated, and traded according to the highest economic value. 'The application of a realistic price policy that reflects underlying scarcities is one of the most efficient ways of ensuring the effective exploitation of a country's resources. In a free economy it is, moreover, the best way of effecting a balance between supply and demand and preventing waste of a scarce commodity. It is, therefore, undesirable that the price of water, either to the urban or to the agricultural sector, be kept artificially depressed through administrative decisions' (DWA 1970).

The importance of private service providers as a stakeholder in water resource management also emerged in this era. Engineering and consultancy firms emerged, who proposed their share in the new niche of bulk water supply. In their view, the role of the state should be limited: 'Except to the extent that it is necessary for security, strategic and practical reasons, the detailed design, construction and maintenance activities should be undertaken by the private sector'. The Department of Water Affairs should restrict itself to undertake 'those functions that are not economically viable or that achieve greater public benefits in the absence of commercial interests' (DWA 1986). DWA acknowledged the challenge: 'To manage and control the services offered by the private sector, which are rapidly increasing in extent and sophistication, the DWA must ensure that it is in a position not only to attract but also to retain a fair share of the finest talent available in the country' (DWA 1986).

In this way, a 'white urban-industrial water economy' emerged out of a specialised private sector, anticipating the business of serving the increasingly wealthier whites who were able and willing to pay for year-round high security water. Water came to be seen as an economic good, the delivery costs of which should be paid for. Moreover, in order to manage demand, it should be allocated to its most profitable use. The fact that the Lesotho Highlands project could obtain loans on a commercial market illustrates the trust of investors that bringing water from hundreds of kilometers to wealthy and risk-averse users is sound business.

While centrally coordinated bulk water provision remained a key task, new planning and regulatory roles for the state were added. Legally, the 1970 Commission adopted the notion of 'assurance of supply' to set priorities, arguing that the new industries and urban water should get their fair share. Also, it recommended that, in certain streams, water should be 'set aside' for such purposes (DWA 1970). Pollution by mines, in particular abandoned mines, urban sewage and agriculture was growing. DWA increasingly addressed these water quality issues by promoting the re-use of urban sewerage effluent to enhance available water quantities, by fixing water quality standards and by monitoring, e.g. through the National Chemical Monitoring Program of the 1970s and the Water Amendment Act, 1984. Research centres like the Water Research Commission, which the 1970 Commission created in 1971, assisted in finding scientific answers to these water quality and other identified challenges.

Lastly, as one of the first countries in the world, the apartheid government started recognising the 'environment' as a water user in its own right, partly for geopolitical reasons. In 1970, 'the environment' was defined as South Africa's two major nature reserves, the Kruger National Park and St. Lucia wetlands. By 1986, this had expanded to include some 300 small, white-controlled public and private nature reserves. Environmental water needs were estimated to require 13% of the total nation's water demand (DWA 1986).

The drastic changes since the 1970s rendered the 1956 Water Act, with its many amendments, outdated. An entirely new Act was needed (DWA 1986).

For water supply and sanitation delivery, no one body was accorded national oversight responsibility, and there was no national policy or standards. Local and regional bodies were responsible for water supply and sanitation delivery. White urban areas were provided with water and sanitation services equal to those of developed countries, while in the black townships, the levels of services varied from full reticulation to standpipes and bucket toilets. Even where reticulated services were provided, levels of operation and maintenance were poor, resulting in high levels of interruptions and leakages.

1.3.3 Entrenching Exclusion

All the above-mentioned new concepts of water management, which accompanied the change from state support to white agriculture to urban economies in often water-scarce areas, were systematically phrased as blanket policies in 'the national interest'. This allowed extrapolating the implications of the agrarian transition in the white economy to all blacks, including those for whom agriculture remained an important mainstay. In the face of the growing political protests and payment boycotts because of weak service delivery, DWA realized that the new criterion of 'economic viability' would not work. It recognised water as a social good: 'Affordability is sometimes an important consideration in the allocation of the full cost of domestic supplies. This is particularly the case for the largest and fastest growing sector of the community, which generally has yet to advance from Third World economic conditions and which, in part, cannot afford the true cost of supplies to meet basic human needs' (DWA 1986).

However, for irrigation in the former homelands, the new norm of 'economic viability' was applied. Arising from the homeland betterment policies of the 1950s some investments had been made in irrigation schemes, with the primary aim of slowing the influx of African men into urban areas. While women were the traditional crop farmers, newly developed irrigation plots were exclusively allocated to men for their supposed full-time 'family farm'. The Tomlinson report stated:

> Out of the various farming and settlement systems, irrigated farming is undoubtedly the enterprise for which the Bantu has proven that they are able, under white management and leadership, to make an economic living out of full-time farming and to use the land economically for food production. Unlike rainfed agriculture, the man does not avoid activities here – the man and his whole family are active on the plots' (Houghton 1956).

Yet women continue to provide most of the labour. Between 70% and 90% or more of irrigating cultivators in, for example, the Olifants river scheme and all other government-supported smallholder irrigation schemes in the Olifants basin, are women. (Makhura and Ngqaleni 1996; Van Koppen 2002). However, women irrigators' control over the fruits of their labour and hence, their incentives to increase productivity, were further curtailed.

The Department of Water Affairs recognised the potential of irrigation. 'Irrigation development is often seen as an important first step in uplifting relatively undeveloped communities, but irrigation development is only one, and not necessarily the most effective, means of achieving socio-economic objectives'. Yet, it now measured with the same yardstick, also protecting white irrigators from market and water competition: 'Since economic advantage is the decisive criterion for successful irrigation, the actual cost of all inputs must be taken into account and proposed projects must be tested against strict efficiency norms'. Moreover, 'Besides the claims made on water of common interest and capital resources, the influence of irrigation on the balance between the demand and supply of agricultural products should also be taken into account' (DWA 1986). In sum, Africans were never meant to become significant water users, let alone 'economically viable' water users who would be able compete with whites, whether on agricultural or water markets. Instead, any new water claim by Africans had now to face yet another new user with stronger rights: 'the environment', as defined by white technocrats.

The 1970s, and especially the 1980s, also saw the growing mass movements against the apartheid regime. The evolving white economy lost all legitimacy. In 1980, when the water planners argued for a high assurance of water supply to SASOL, this very symbol of the white economy was blown up by the armed wing of the underground military wing of the ANC (Saunders and Southey 2001). National and international protests and boycotts against the racist apartheid regime made the country ungovernable and led to the negotiated settlement for democracy.

1.4 Water Reform in the Context of National Transformation

Under the new dispensation post 1994, the overarching mission of the new government, including the Department of Water Affairs and Forestry, changed from serving a small formal and well-organised white constituency to serving an entire nation of over 40 million citizens, deeply divided by wealth differences and territorial and institutional segregation. In 1996, a new Constitution (Act 108 of 1996) (RSA 1996) was adopted, which enshrined equal rights for all. The Constitution had important implications for water services provision in South Africa. The two key Constitutional elements were the assignment of the water services provision function to municipalities and the right, set out in the Bill of Rights, of access to sufficient water for all in South Africa. The right to an environment, not harmful to health or well-being, as enshrined in the Bill of Rights is also important.

The structure of the state also changed. The despised homeland government structures were dismantled, although tribal chiefs, the organs through which the homeland governments had worked on the ground, maintained strong powers, in particular over communal land tenure. A new unitary national government structure was set up. Provincial Government was established in nine new provinces. In 2000, 284 new municipalities were demarcated that covered the entire country. Various pieces of legislation were developed to guide municipalities in the execution of their constitutional function to provide water supply and sanitation, such as the Water Services Act (Act 108 of 1997), the Local Government Municipal Demarcation Act 27 of 1998, the Municipal Structures Act 117 of 1998, the Municipal Structures Amendment Act 33 of 2000, and the Municipal Systems Act 32 of 2000. Many municipalities crossed former homeland and new provincial boundaries. In the former homelands, local government was established alongside, and often at loggerheads with, tribal authorities. In former white areas, boundaries changed, often to include large poor "black" areas.

The new government also launched land reform with three components: restitution of land that was appropriated by whites after 1913; redistribution of land in former white areas; and tenure reform for tenants and in former homelands. By the early 2000s, significant portions of white areas were under claim (Lahiff 2007).

Prof. Kader Asmal, Minister of Water Affairs and Forestry from 1994–1999, created an unprecedented momentum of intense and wide-ranging debate that tapped extensive national expertise and experiences from high- and middle-income countries, to draft the National Water Act (RSA 1998). The Act, which was promulgated in 1998, addresses the double challenge of, on the one hand, maintaining what was good from the previous dispensation, in particular the continuing operation of the massive asset base of infrastructure and government's growing regulatory role for large-scale users, and, on the other hand, changing the wrongs of the past. The latter is clearly the greatest challenge – fundamentally changing access to water for productive purposes by black South Africans.

The aims of the National Water Act are, amongst others:

- To meet the basic human needs of present and future generations
- To promote equitable access to water
- To redress the results of past racial and gender discrimination
- To promote the efficient, sustainable, and beneficial use of water in the public interest
- To facilitate social and economic development

[...] and to achieve this purpose, to establish suitable institutions and ensure that they have appropriate community, racial and gender representation (NWA, Chapter 1, Section 2). The Act has various components, the operationalisation and implementation of which are discussed in following chapters.

Within DWAF itself, after 1994, the race and gender composition became considerably more representative (Garduño and Hinsch 2005). Many former white male staff left DWAF voluntarily, for retirement or to join the better-paying private consultancy firms that offered their services, among others, to DWAF. Affirmative action contributed to this more representative staff composition. However, high turn-over among black employees departing to the better-paying private sector or for rapid promotion in government, as well as limited technical skills in the country, contributed to concerns about technical capacity with DWAF and, more broadly, in the sector.

1.5 Structure of the Book

What has been described above is the context in which the measures discussed in this volume were taken. With regard to the state's role in water management, the central planning of raw water supply to a water economy that was being transformed from being white only was continued. The overview of available water resources demands and supplies throughout the newly declared Water Management Areas in the country, as presented in Chapter 2, synthesised decades of institutional memory and recent updates to inform centralised water resources management in South Africa.

Alongside to this, a national programme for the delivery of water services to all was put in place. As discussed in Chapter 3, this programme was intended to meet the enormous backlogs in domestic water services to black South Africans, particularly in rural areas. The White Paper on Community Water Supply and Sanitation of 1994 provided the framework for delivery of basic services, which led to the Water Services Act of 1997 (RSA 1997). By 2000, municipalities had been formally established and their powers and functions assigned, and the process of transfer of schemes and the backlog delivery programme to municipalities began. In 2003, the Strategic Framework for Water Services was adopted by Cabinet (RSA 2003). The Strategic Framework recognises the role of municipalities in providing sustainable water services, and the changed role of the Department of Water Affairs and Forestry as support, monitoring and regulation rather than direct service provision.

Chapter 4 deals with the challenges of providing sanitation in the urban context of contemporary South Africa and highlights some of the key lessons from this area of work. The provision of sustainable and adequate sanitation remains a significant challenge, made more difficult by the race and wealth divisions in South African society. The chapter highlights that the state's ability to operate and maintain infrastructure does not match its ability to deliver new infrastructure to the previously unserved. This poses significant challenges in the delivery of sanitation, and brings significant risks from poorly maintained and managed waste water treatment works.

The legal changes introduced by the NWA and their implementation are discussed in Chapter 5. The NWA replaced the water entitlements system of the pre-1994 patchwork of common-law riparian rights, groundwater as private property in white areas, poorly-defined customary water rights regimes in former homelands, irrigation scheme arrangements, state authority to set priority assurances of supply, and a number of Government Water Control Areas. Instead, a civil-law water rights system was adopted, in which the state is Trustee of all the nation's surface *and* ground water resources, with the responsibility to authorise 11 identified uses. This revolutionary move, which in effect, nationalised water resources, allowed a stronger regulatory role for DWAF to ensure that water allocation redressed the results of past racial and gender discrimination (RSA 1998).

Chapter 5 presents the lessons learnt from the early pilot implementations of Compulsory Licensing, focusing on whether and how it, ultimately, contributes to redressing race- and gender-inequities. In the 12 years since 1994, very little land or water has actually changed hands, due to the slow pace of land and water reform. The same is true for new water uptake – of the 1,200 or so new licences issued between 1998 and 2006, 98% were for white users. These experiences raise growing concerns about the appropriateness of formal licence systems for reaching millions of scattered informal small-scale users, who are typically, considerably less administration-proficient than vested large-scale users. Forms of recognising informal water rights regimes, General Authorisations and prioritisation measures are emerging as remedies (DWAF 2006b; Van Koppen et al 2007).

In the new water legislation, 'the environment' and water for basic human needs were given the highest allocation priority. The Reserve is the only water *right* in South Africa, and covers both ecological requirements and basic human needs. The latter echoes the most universal water allocation principle across the world, that nobody can deny people's access to water for drinking and basic domestic needs. Chapter 6 analyses the ecological elements of the Reserve and other policy approaches to protecting aquatic systems. The work done by South Africa in this area has given rise to strong inputs into global debates on priority allocation for environmental flows.

Another innovation in the National Water Act that attracted much international attention concerns new water governance structures, as discussed in Chapter 7. The Act enables the establishment of Catchment Management Agencies (CMAs). In performing their functions, CMAs must be 'mindful of the constitutional imperative to redress past racial and gender discrimination and to achieve equitable access for all to water resources under its control' (NWA, Chapter 7, Section 79–4a). CMAs are seen as a critical part of democratising water management in South Africa. Immediately after promulgation of the Act, 19 Water Management Areas were demarcated for 19 CMAs, although this is under review. Until the establishment of functioning CMAs, DWAF acts as the CMA.

For localised water management in a mutual interest, the Act enables water users to organise into Water User Associations and requires the transformation of previously white Irrigation Boards into Water User Associations to ensure racial and gender equity.

Chapter 7 analyses the implementation of the institutional components of the Act and their alignment with other water institutions, such as Water Boards, and more recent actors in the institutional landscape, namely provincial and local government.

There were also innovations in national planning and bulk water supply delivery, as elaborated in Chapter 8. The NWA prescribes national planning with considerable public participation through the National Water Resources Strategy (NWRS). Lower-level Catchment Management Strategies should 'not be in conflict' with the NWRS. In compiling its first strategy, DWAF devoted major resources to realising broad public participation from 2002 to 2004.

The operation and maintenance of the bulk water supply assets of the past, to meet the water needs of the gradually less race-based economically viable water users, continued after 1994. Generally, plans for new infrastructure that were already in motion in the 1980s were implemented. Some plans were accelerated, such as bulk domestic supplies to the formerly unserved, whose right to domestic water was pursued as a top priority. Other plans were delayed, largely because of international and local concern with the impacts of large dams and the call for procedures that considered social and environmental issues. As chair of the World Commission on Dams, Minister Kader Asmal encouraged debate and innovation from the very diverging views within and outside South Africa.

As a result of these debates, public participation, proper compensation to those displaced, and Environmental Impact Assessments are now an integral part of dam development. Internationally, DWAF asserted the importance of increased water storage, in developing countries, to ensure multi-year storage as a vital component of IWRM.

Chapter 9 discusses the financial conditions for water supply. The National Water Act incorporates the 'user pays' principle, first raised in the 1970s. As a result, subsidies to white farmers were ended and a pricing strategy for raw water was introduced that included a charge of infrastructure and a much smaller charge to cover the costs of water resources management. Waste discharge charges were introduced into the strategy later. When implemented, income from the latter is expected both to provide a large part of the incomes for CMAs and to control pollution effectively.

Chapter 9 also discusses DWAF's subsidy program for resource-poor farmers (DWAF 2004). This subsidy is based on the recognition that active intervention by the state is required to assist black farmers to overcome the legacy of apartheid exclusion.

Chapter 10 discusses the remarkable changes achieved by and for women, both white and black, in terms of gender and race composition and decision-making within DWAF. Improved access to water for domestic purposes, as discussed in Chapter 3, benefited women in relation to their domestic chores more than in relation to productive activities. Especially in the former homelands, access to water for women's and men's productive uses has generally reduced. With the dismantling of the apartheid structures, support to smallholder irrigation schemes was withdrawn. Schemes deteriorated or even collapsed (Shah et al 2002). As discussed in this chapter, the design of alternative integrated support for financing, technical options and institutional arrangements, appeared a tall order. Many sectors are involved in productive water uses. In the past, the small white professional community could smoothly co-operate across government departments to deliver the integrated support package needed for the interests of the broader apartheid economy. Under the new dispensation, however, the forging of structures and relations to do so in a democratic manner are still underway.

Chapter 11 discusses the challenges pertaining to information systems in supporting the new approaches to water resources management being implemented in South Africa. The research contribution to South Africa's water management challenges are discussed in Chapter 12.

Chapter 13 deals with innovations in transboundary water management arrangements. Under the National Water Act, international obligations are second only to providing water for basic human needs and for ecological functioning. The commitment to ensuring water for international purposes is set within the context of the SADC Revised Protocol on Shared Watercourses, which is one of the few regional agreements on the shared management of transboundary river basins.

The final chapter in the volume draws lessons from the preceding chapters, under a framework of good water governance. The framework draws on three key ingredients for good water governance: clear policy, a strong bureaucracy and an informed civil society.

Acknowledgement This chapter was prepared with the support of PN17 "Integrated water resources management for improved rural livelihoods", a project of the CGIAR Challenge Program on Water and Food.

References

Backeberg GR (1994) Die politieke ekonomie van besproeiingsbeleid in Suid-Afrika. Voregle ter vervulling van 'n deel van die vereistes vir die graad Philosophiae Doctor. Departement Landbou-ekonomie. Voorligting en Lndelijke Ontwikkeling. Fakulteit Biologiese en Landbouwetenskappe. Universiteit van Pretoria, Pretoria

Bundy C (1988) The rise and fall of the South African peasantry, 2nd edn. David Philip, Cape Town/Johannesburg

Crais C (1992) The making of the colonial order: white supremacy and black resistance in the Eastern Cape. Witwatersrand University Press, Johannesburg, pp 1770–1865

Cullis J, van Koppen B (2007) Applying the Gini coefficient to measure the inequality of water use in the Olifants river water management area, South Africa. IWMI Research Report 113. International Water Management Institute, Colombo
Delius P (1983) The land belongs to us: the Pedi polity, the Boers, and the British in the nineteenth-century Transvaal. Ravan Press, Johannesburg
DWA (Department of Water Affairs) (1956) Water act. Republic of South Africa, Pretoria
DWA (Department of Water Affairs) (1970) Report of the Commission of Enquiry into water matters. R.P. 34/1970. Republic of South Africa, Pretoria
DWA (Department of Water Affairs) (1986) Management of the water resources of the Republic of South Africa. Department of Water Affairs, Pretoria
DWAF (Department of Water Affairs & Forestry) (2004) Policy on financial assistance to resource-poor irrigation farmers in terms of Sections 61 and 62 of the National Water Act (1998). Department of Water Affairs & Forestry, Pretoria
DWAF (Department of Water Affairs & Forestry) (2006b) Guidelines for water allocation in South Africa, a framework for water allocation reform. Department of Water Affairs and Forestry, Pretoria
Fényes T, Meyer N (2003) Structure and production in South African agriculture. In: Nieuwoudt L, Groenwald J (eds) The challenge of change. Agriculture, land and the South African economy. University of Natal Press, Pietermaritzburg
Garduño H, Hinsch M (2005) IWRM Implementation in South Africa redressing past inequities and sustaining development with a view to the future (Reporting to: Ana María Kleymeyer). World Bank Institute, Washington, DC
Houghton DH (1956) The Tomlinson report. A summary of the findings and recommendations in the Tomlinson Commission Report. The South African Institute of Race Relations, Johannesburg
Lahiff EP (1999) Land tenure on the Arabie-Olifants irrigation scheme. South Africa. Working Paper 2. Nkuzi Development Association and International Water Management Institute. International Water Management Institute, Colombo
Lahiff E (2007) State, market or the worst of both? Experimenting with market-based land reform in South Africa. Land reform and agrarian change in South Africa. Occasional papers series No. 30. Programme for Land and Agrarian Studies. Western Cape School of Government, University of Western Cape, Cape Town
Lewis AD (1934) Water law. Its development in the Union of South Africa. Juta and Co. Ltd, Cape Town/Johannesburg
Makhura T, Ngqaleni MT (1996) An analysis of women's status in agricultural development in the Northern Province. In: Lipton M, Ellis F, Lipton M (eds) Land, labour and livelihoods in Rural South Africa, vol 2. Indicator Press, Durban
Monnich HO (1967) The Pedi. J.L. van Schaik Ltd., Pretoria
O'Meara D (1983) Volkskapitalisme. Class, capital and ideology in the development of Afrikaner nationalism. Ravan Press, Johannesburg, pp 193–1948
RSA (1994) White paper on water supply and sanitation. Government Printer, Pretoria, South Africa
RSA (1996) The Constitution of the republic of South Africa (Act No. 108 of 1996). Government Printer, Pretoria
RSA (1997a) White paper on a national water policy for South Africa. Pretoria, Government Printer
RSA (1997b) Water Services Act. Act 108 of 1997. Government Gazette No. 18522. Office of the President, Cape Town
RSA (1998, August 26) National Water Act. Government Gazette Vol. 398. Office of the President, Cape Town
RSA (2003) Strategic framework for water services. Government Printer, Pretoria, South Africa
Saunders C, Southey N (2001) A dictionary of South African history. David Philip, Cape Town/ Johannesburg
Shah T, van Koppen B, Merrey D, de Lange M, Samad M (2002) Institutional alternatives in African smallholder irrigation: lessons from international experience with irrigation management transfer. IWMI Research Report No 60. IWMI, Colombo

SSA (2000) Employment trends in agriculture in South Africa. A joint publication between Stats SA and the National Department of Agriculture. Statistics South Africa, Pretoria

Terreblanche S (2002) A history of inequality in South Africa 1652–2002. University of Natal Press, Pietermaritzburg

Thompson L (2001) A history of South Africa. Yale Note Bene Book. Yale University Press, New Haven/London

Turton AR, Meissner R (2003) A hydropolitical history of South Africa's international river basins. Report to the Water Research Commission. African Water Research Issues Unit, University of Pretoria, Pretoria, South Africa

Van Koppen B (2002) A gender performance indicator for irrigation: concepts, tools, and applications. Research Report 59. International Water Management Institute, Colombo

Van Koppen B, Giordano M, Butterworth J (2007) Community-based water law and water resource management reform in developing countries. Comprehensive Assessment of Water Management in Agriculture Series 5. CABI, Wallingford

Vink N, Kirsten J (2003) Agriculture in the national economy. In: Nieuwoudt L, Groenwald J (eds) The challenge of change. Agriculture, land and the South African economy. University of Natal Press, Pietermaritzburg

Chapter 2
Water Resource Situation, Strategies and Allocation Regimes in South Africa

Johan van Rooyen, Marna de Lange, and Rashid Hassan

Abstract This chapter provides background to the state of water resources in South Africa and how those resources have been strategically developed and managed and the policy environment that guided their allocation and use pre- and post-1994. The topography and weather patterns strongly influence rainfall and evaporation, which result in highly varying rainfall patterns and even more erratic surface run-off. Requirements for water and the resultant needs for the development of the resource have also been strongly influenced by social, political and economic factors. The chapter gives a brief account of the water supply and management infrastructure that was developed. The current overall balance between available water and water requirements is also analysed as well as the challenges that water resource managers in South Africa are currently facing in a country that is rapidly approaching the limits of its available fresh water supplies.

Keywords Water allocation • Planning • Water scarcity • Shared river basins • Infrastructure • Yield • Subsidies

---

J. van Rooyen (✉)
Directorate: National Water Resource Planning, Department of Water Affairs, Pretoria, South Africa
e-mail: VanRooyenJA@dwa.gov.za

M. de Lange
Socio-Technical Interfacing Consulting CC, Polokwane, South Africa
e-mail: marna@global.co.za

R. Hassan
Centre for Environmental Economics and Policy in Africa, South Africa
e-mail: rashid.hassan@up.ac.za

B. Schreiner and R. Hassan (eds.), *Transforming Water Management in South Africa: Designing and Implementing a New Policy Framework*, Global Issues in Water Policy 2, DOI 10.1007/978-90-481-9367-7_2,

2.1 Introduction

This chapter provides background to the state of water resources in South Africa and how those resources have been strategically developed and managed and the policy environment that guided their allocation and use pre- and post-1994. The next section shows how the topography and weather patterns influenced rainfall and evaporation, which result in highly varying rainfall patterns and even more erratic surface run-off. Section 2.3 provides an overview of how the social, political and economic factors influenced the requirements for water and the resultant needs for the development of the resource. Section 2.4 gives a brief account of the water supply and management infrastructure that was developed. The current overall balance between available water and water requirements is analysed in Section 2.5 and the final section concludes the chapter identifying the challenges water resource managers in South Africa are currently facing in a country that is rapidly approaching the limits of its available fresh water supplies.

2.2 Physical and Hydrological Conditions

South Africa is situated in the high pressure belt of the middle latitudes of the Southern Hemisphere. High pressure systems tend to dominate the weather over large parts of the country and this is unfavourable for the formation of rain. The weather is influenced by the warm Agulhas current that flows along the east coast and the cold Benguela current flowing along the west coast of the country. The topography of the country also has a major influence on rainfall patterns with the land rising steeply from the eastern and southern coastline to mountains forming the rim of the large interior plateau and then falling gradually to the north and west (DWA 1986).

The interaction between these factors results in complex weather and rainfall patterns over the country. Rainfall generally decreases in a westerly and northerly direction as shown in Fig. 2.1. About 65% of the country receives less than 500 mm of rainfall on average per year, which is regarded as the limit for successful dry-land cultivation of crops. About 20% receives less than 200 mm per year with desert conditions on the far western parts (DWA 1986).

This spatial variation in rainfall is further complicated by large variations over time. Temporal variations tend to follow the same pattern with the least variation between years on the eastern and southern coastal areas and the largest variation towards the dry western areas as can be seen in Fig. 2.2 (DWA 1986).

The country is predominantly a summer rainfall area, but with a winter rainfall area on the west coast and a small band along the south coast where rain falls throughout the year. Evaporation is high, varying from a potential of around 1,500 mm per year in the east and south to as high as 3,000 mm per year in the western desert areas as shown in Fig. 2.1.

The large variations in rainfall result in even larger variations in river flows. This is because the soil moisture must be replenished before significant run-off

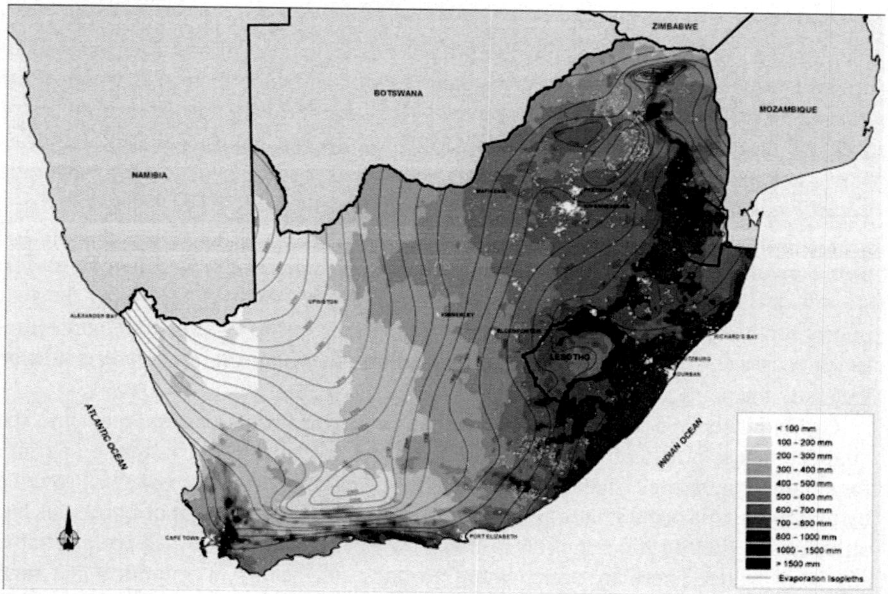

Fig. 2.1 Rainfall and evaporation in South Africa

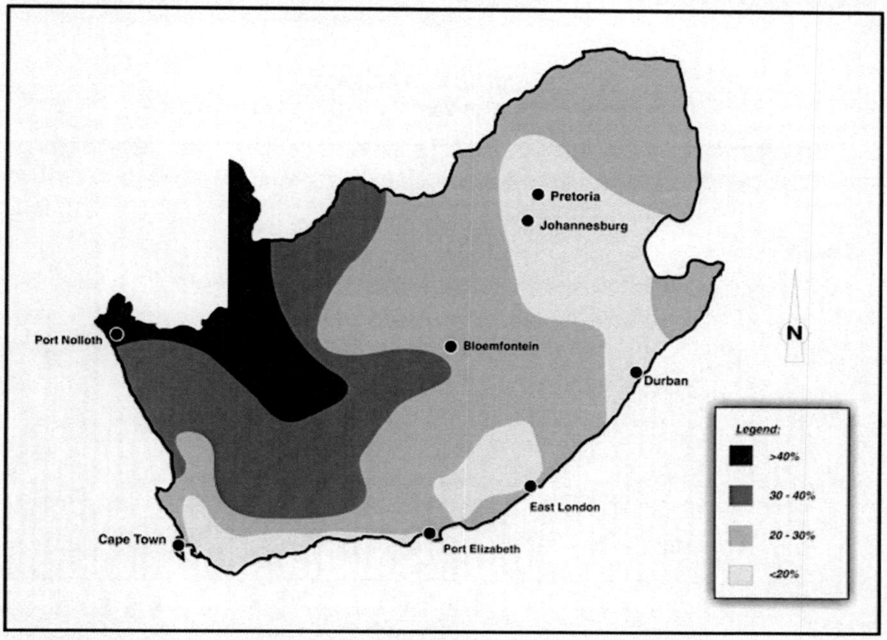

Fig. 2.2 Percentage deviation from mean annual rainfall

occurs. If rainfall events are small or irregular the large evapotranspiration and evaporation deplete soil moisture before river flow can occur and long periods of drought ensue. This is very clearly illustrated by the rainfall and runoff statistics of the Vaal River at Vaal Dam in Figs. 2.3 and 2.4. The rainfall varies from a minimum of 55% of the mean to a maximum of 152% while the naturalised runoff varies between 15% and 370%. On the rainfall graph the longest period of below average rainfall was 5 years but for the run-off it was 9 years (SANCOLD 1994).

The large variations in the flow to rivers invariably mean that storage is required to make sufficient quantities of water available for users at times of low flow. The storage in dams is required not only to store water from the wet months of the year in order to be used in the months when river flow decreases dramatically and sometimes ceases to flow, but also over drought periods where the river flow could be very low for many consecutive years.

Groundwater in South Africa occurs mainly in hard rock aquifers that limit the quantity that is practically available for use. It is a very important source of water for small communities because the storage that would otherwise have to be provided by building dams occurs naturally underground. Poor management of boreholes has unfortunately led to a perception in many cases that groundwater is not a reliable source of water. There are areas where groundwater occurs in dolomitic and sand aquifers and in those areas large volumes of water are abstracted, currently mainly for irrigation purposes. The deep fracturing in the sandstones of the Cape Fold mountains has been identified as a possible source of large quantities of groundwater that has not been utilised up to now. It has the potential of supplying large volumes

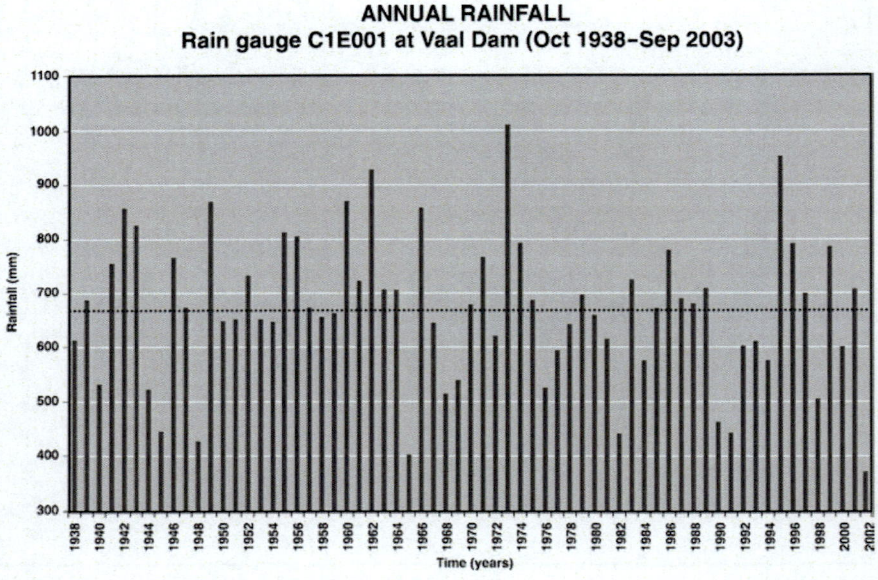

Fig. 2.3 Annual rainfall at Vaal Dam

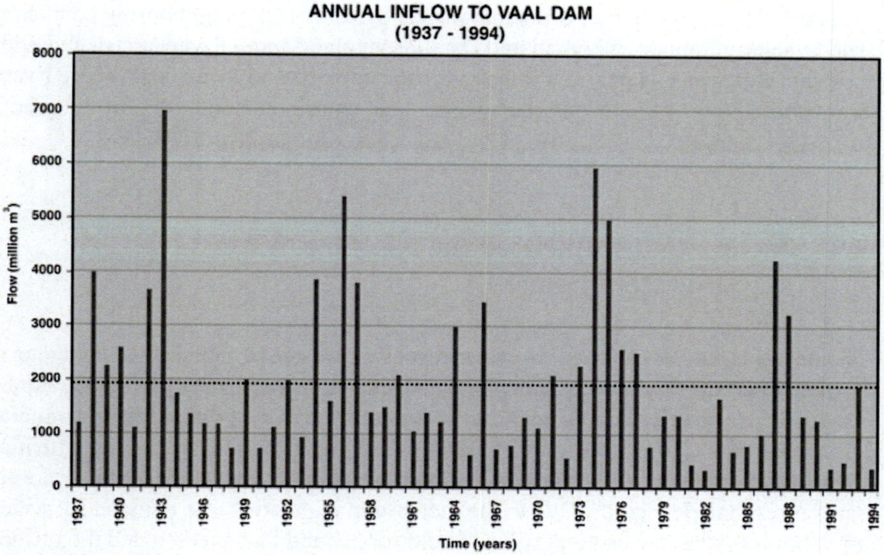

Fig. 2.4 Annual inflow into Vaal Dam

of water that may be a significant resource for some of the metropolitan areas along the Cape coast (DWA 1986; DWAF 2004).

The quality of surface water under natural conditions is generally good in South Africa except for a number of cases where, due to natural geological conditions, the water would become salty under low flow conditions. That is also reflected in the names of rivers like a number of "Salt Rivers" and "Brak Rivers".

However, man has had a significant impact on the quality of the water with return flows or effluent from urban and industrial areas. The large gold and coal mining areas of the country has also caused major water quality problems. The large areas of land under cultivation, both for rain fed and irrigated agriculture and the associated use of fertiliser and pesticide is also a contributor to water quality deterioration. Depending on the specific source the problems vary from bacteriological contamination, salts with sulphates specific problem, nutrients and pesticides and other dangerous chemicals (DWAF 2004).

The quality of groundwater varies dramatically under natural conditions, again depending on the geological conditions with very high salt contents and high fluoride concentrations rendering water unfit for use in some areas. Human activity has also impacted seriously on some groundwater resources in the country. Such impacts on the groundwater are even more serious than surface water pollution as it is very difficult and costly to rehabilitate and will take a very long time to achieve.

All the above suggests that South Africa is a water scarce country with complex hydrological conditions which makes it very challenging to secure sufficient water for economic growth and social needs in a sustainable manner.

South Africa shares four of its major river systems with neighbouring countries. The Orange, Limpopo, Nkomati and Usutu/Pongola/Maputo Rivers are shared with Lesotho, Botswana, Namibia, Zimbabwe, Mozambique and Swaziland. These rivers drain about 60% of South Africa's surface area, contribute about 40% of the runoff and support approximately 70% of the gross domestic product (DWAF 2004).

2.3 Factors that Influenced the Need for Water Resource Development

The need for the development of water resources was determined by many factors. As the population of South Africa grew and spread, towns and villages were established. This normally happened close to a reliable source of water like a perennial spring or stream. Towns with favourable locations and good water had the potential to grow and some of these towns and villages became important economic or political centres in their regions. Due to the increase in population, the demand for water grew beyond what the original source could supply and that necessitated the further development of water resources. One example of this is Bloemfontein, whose name indicates the existence of a beautiful fountain.

The town became the capital of the Republic of the Free State and is still the capital of that province. Dams were built in the rivers close to the town, but the requirements have grown to such an extent that water now has to be transferred over long distances from large dams in a neigbouring catchment to supply the requirements. Another example is Pretoria, with its original source of water the dolomitic fountains in the area. Pretoria became the capital of the Transvaal Republic and later the capital of South Africa and some of its water is now transferred from the Lesotho Highlands Water Project (DWA 1986; DWAF 1997, 2004).

Along the coast the sites for good harbours often dictated the establishment of the larger coastal towns and cities. South Africa lacks large rivers that could be used for harbours and the favourable harbour sites are often far away from large water resources.

The mineral deposits of the country are another main factor that determined the settlement patterns and in the majority of cases were far away from the major water resources. A good example is the gold fields of the Witwatersrand that led to the establishment of Johannesburg around which the biggest metropolitan area of the country eventually developed. The Witwatersrand is a series of hills that form the continental divide in this area. The water draining to the north flows down the Limpopo River system to the Indian Ocean, while the water draining to the south flows down the Vaal and Orange River system to the Atlantic Ocean. This area is about as far away as one can get from a large water resource (DWA 1986; DWAF 2004).

Irrigation development was another main driver of water resource development in South Africa. Under Dutch rule and the Roman Dutch Law the State had full control over the use of water in rivers in the Cape Colony. This changed under the British rule where, under English Law, the right to use of water was given to riparian owners.

Irrigation at this stage was mostly done from small private run-of-river schemes with small canals taking water from the rivers to the irrigated fields (DWA 1986).

In 1912 the Irrigation and Conservation of Waters Act was promulgated and played a huge role in promoting irrigation development. The economic depression after the First World War became an important motivation for the development of irrigation schemes. This trend accelerated with the great economic depression in 1929 and a number of large Government irrigation schemes were built. Destitute white people from drought stricken areas were employed to build the labour intensive schemes and many of them were then settled on these schemes as irrigation farmers (DWA 1986).

After the Second World War it was realised that the 1912 Act was no longer suitable for the more industrialised phase that the economy of SA had entered and a Commission of Enquiry was appointed to investigate matters. The Commission reported in 1952 and this led to the promulgation of the Water Act (Act 54 of 1956) (DWA 1956). The Act recognised the principle of private ownership and water court awards, but also introduced the concept of State control. The State could then control water use that was in excess of existing rights, if necessary. At the same time the Department of Irrigation was renamed to the Department of Water Affairs.

Almost at the same time that a greater appreciation for the role of water in the industrial development of the country took hold, the political development took a turn for the worse with the National Party government strengthening the "apartheid" policies. This eventually led to the establishment of so-called self-governing areas as well as the four nominally independent "homeland" states of Venda, Transkei, Ciskei and Bophuthatswana. As a result, water resource management and development became highly fragmented with each of these areas responsible for their own water resources management.

These territories were very rural in nature and economic development almost non-existent. The requirement for water was consequently also very low and even water supply for basic human needs very much neglected.

In some of these areas there was a drive by the "governments" to develop irrigation schemes and dams were built to make this possible. For many reasons these developments were not sustainable and the schemes fell into disuse as soon as government assistance (both monetary and managerial) was withdrawn after 1994.

2.3.1 The Infrastructure that was Developed

The factors that determined the development of water resources are described in the previous paragraphs. A large number of dams, from very small to very big, were built to make water available. The dams for irrigation purposes were mainly developed as "stand alone" dams and operated individually. The larger dams would also very often have canal and pumping stations and pipeline systems associated with it to distribute the water to the irrigation lands.

The infrastructure to supply the large metropolitan areas grew from simple "stand alone" dams into very complex supply systems consisting of a number of large dams, tunnels, pumping stations, pipelines and interbasin transfer schemes. Since the promulgation of the Water Act of 1956 the development of these systems were done by the State in the form of Government Water Schemes (SANCOLD 1994).

A good example of such a system is the Vaal River system that developed over a period of more than 90 years. It started with the building of the Vaal Barrage in 1922 with a pumping station and pipeline to supply the growing needs of Johannesburg. This system currently sources its water from seven large catchments, and supplies water to users in 6 provinces. The system now consists of at least 30 large dams, hundreds of kilometres of large diameter pipelines, tunnels, canals and pumping stations. About 45% of the population of South Africa lives in its supply area and the area is responsible for 75% of the GDP. Users include the urban sector with domestic and industrial users, gold, diamond and coal mines, thermal power stations that uses the water for steam generation and a large quantity for cooling, as well as irrigated agriculture. Only one of South Africa's nine provinces is not impacted on by this huge system, either as a source or receiver of the water (DWAF 2004).

Similar systems, although not as extensive, have been developed for the Western Cape system supplying the metropolitan areas of the Western Cape Province around the Cape Town metropole; the Algoa system supplying the Nelson Mandela Bay area around Port Elizabeth in the Eastern Cape Province; the Buffalo system supplying the East London area in the same province; the Mgeni System supplying the areas around Durban and Pietermaritzburg and the Mhlatuze system supplying the industrial area around the port of Richards Bay in KwaZulu-Natal, as well as the system supplying the metropolitan area of Bloemfontein in the Free State Province.

2.4 Current Water Balances and Implications for Water Resource Management, Quantity and Quality Aspects

The Department of Water Affairs and Forestry has gone through an extensive exercise to calculate water balances for all the water management areas in the country for the year 2000 and has published this in the National Water Resource Strategy. There are many ways of calculating a water balance, but in the NWRS the balance was calculated between the yield of the resource and the use from the resource, all at a 1:50 year assurance of supply. The actual average use from the resource may be considerably more than this, but for a fair balance, it must be compared with the available resource at the same assurance of supply.

Water balances at the water management area scale could be very misleading and careful interpretation is always important. A deficit at the outlet of a large catchment invariably means that the catchment is under stress. It is possible that some of the sub-catchments may have a positive balance or a surplus, but if a further allocation is made from that sub-catchment, the shortage at the outlet will be increased and the overall situation will deteriorate. The converse is also true. If the large catchment has a surplus at the bottom it does not mean that all the sub-catchments upstream are in surplus. It is theoretically possible that all sub-catchments upstream but one may be in deficit, with the one supplying enough of a surplus to show a surplus at the bottom.

The following tables are all from the National Water Resources Strategy (DWAF 2004). Table 2.1 shows the mean annual runoff and estimates of the volume of water that would be required to remain in the rivers in order to meet the requirement of the ecological Reserve in compliance with the National Water Act (Act 36 of 1998). It also shows the storage volume of the large dams in the water management areas.

Table 2.2 shows the estimates for the available yield made up from surface and ground water as well as from the usable return flows, all at a 1:50 year assurance of supply.

Table 2.1 Natural mean annual runoff and the ecological Reserve (million m³/a) and storage in major dams (million m³)

Water management area		Natural mean annual runoff[a]	Ecological reserve[a,b]	Storage in major dams[c]
1	Limpopo	986	156	319
2	Luvuvhu/Letaba	1,185	224	531
3	Crocodile West and Marico	855	164	854
4	Olifants	2,040	460	1,078
5	Inkomati[d]	3,539	1,008	768
6	Usutu to Mhlatuze[e]	4,780	1,192	3,692
7	Thukela	3,799	859	1,125
8	Upper Vaal	2,423	299	5,725
9	Middle Vaal	888	109	467
10	Lower Vaal	181	49	1,375
11	Mvoti to Umzimkulu	4,798	1,160	827
12	Mzimvubu to Keiskamma	7,241	1,122	1,115
13	Upper Orange[f]	6,981	1,349	11,711
14	Lower Orange[g]	502	69	298
15	Fish to Tsitsikamma	2,154	243	739
16	Gouritz	1,679	325	301
17	Olifants/Doring	1,108	156	132
18	Breede	2,472	384	1,060
19	Berg	1,429	217	295
	Total for South Africa	49,040	9,545	32,412

Source: DWAF 2004

[a] Quantities refer to the water management area under consideration only, thus water originating in that water management area

[b] Total volume, based on preliminary estimates. The impact on yield will be smaller than this number

[c] Includes dams constructed to end-2003 with capacity exceeding 1 million/m³, but excludes dams constructed solely for flood control (Qedusizi in WMA 7 and Beervlei in WMA 15). Accounts for accumulated sediment according to most recent surveys and includes dead storage

[d] Includes the Komati catchment in Swaziland (mean annual runoff = 517 million m³/a)

[e] Includes the Pongola catchment in Swaziland (mean annual runoff = 213 million m³/a)

[f] Storage includes Katse and Mohale dams in Lesotho

[g] Includes contributions from the Senqu and Caledon Rivers in Lesotho (mean annual run-off = 4 765 million m³/a).

Table 2.2 Available yield in year 2000 (million m³/a)

Water management area		Natural resource: Surface water[a]	Natural resource: Ground water[b]	Usable return flow: Irrigation	Usable return flow: Urban	Mining and bulk industrial	Total local yield
1	Limpopo	160	98	8	15	0	281
2	Luvuvhu/Letaba	244	43	19	4	0	310
3	Crocodile West and Marico	203	146	44	282	41	716
4	Olifants	410	99	44	42	14	609
5	Inkomati	816	9	53	8	11	897
6	Usutu to Mhlatuze	1,019	39	42	9	1	1,110
7	Thukela	666	15	23	24	9	737
8	Upper Vaal	598	32	11	343	146	1,130
9	Middle Vaal	(67)	54	16	29	18	50
10	Lower Vaal[b]	(54)	126	52	0	2	126
11	Mvoti to Umzimkulu	433	6	21	57	6	523
12	Mzimvubu to Keiskamma	777	21	17	39	0	854
13	Upper Orange	4,311	65	34	37	0	4,447
14	Lower Orange[c]	(1,083)	24	96	1	0	(962)
15	Fish to Tsitsikamma	260	36	103	19	0	418
16	Gouritz	191	64	8	6	6	275
17	Olifants/Doring	266	45	22	2	0	335
18	Breede	687	109	54	16	0	866
19	Berg	403	57	11	37	0	505
	Total for country	10,240	1,088	675	970	254	13,227

Source: DWAF 2004

Transfers into and out of water management areas are not included above, but are covered in Table 2.4

[a] Yield from run-of-river and existing storage, after allowance for the impacts on yield of the ecological component of the Reserve, river losses, alien vegetation, rain-fed sugar cane and urban runoff

[b] Estimated use from existing boreholes ad springs. As a result of development of groundwater for irrigation since the compilation of the database for the NWRS, total groundwater use may exceed this estimate. The increase is mainly due to growth in irrigation water requirements, and therefore does not significantly impact on the overall water balances given in the NWRS

[c] Negative figures under surface water caused by river losses being larger than the incremental runoff from within the water management area

In comparing Table 2.2 with Table 2.1 it can be seen that the available surface water yield is about 20% of the mean annual runoff. Due to the erratic runoff the bulk of this yield is supplied from the large dams.

Table 2.3 gives the water requirement for the year 2000. Water for irrigation makes up 62% of the requirement and urban use 23%, with the rest for rural, mining and industries outside urban areas, power generation and afforestation.

Table 2.4 gives the balance or reconciliation of the requirements and the yield as at 2000. This table shows that 10 of the 19 water management areas were already

Table 2.3 Water requirements for the year 2000 (million m^3/a)

Water management area		Irrigation	Urban[a]	Rural[a]	Mining and bulk industrial[b]	Power generation[c]	Afforestation[d]	Total local requirements
1	Limpopo	238	34	28	14	7	1	322
2	Luvuvhu/Letaba	248	10	31	1	0	43	333
3	Crocodile West and Marico	445	547	37	127	28	0	1,184
4	Olifants	557	88	44	94	181	3	967
5	Inkomati	593	63	26	24	0	138	844
6	Usutu to Mhlathuze	432	50	40	91	0	104	717
7	Thukela	204	52	31	46	1	0	334
8	Upper Vaal	114	635	43	173	80	0	1,045
9	Middle Vaal	159	93	32	85	0	0	369
10	Lower Vaal	525	68	44	6	0	0	643
11	Mvoti to Umzimkulu	207	408	44	74	0	65	798
12	Mzimvubu to Keiskamma	190	99	39	0	0	46	374
13	Upper Orange	780	126	60	2	0	0	968
14	Lower Orange	977	25	17	9	0	0	1 028
15	Fish to Tsitsikamma	763	112	16	0	0	7	898
16	Gouritz	254	52	11	6	0	14	337
17	Olifants/Doring	356	7	6	3	0	1	373
18	Breede	577	39	11	0	0	6	633
19	Berg	301	389	14	0	0	0	704
	Total for country	7 920	2 897	574	755	297	428	12,871
		62%	23%	4%	6%	2%	3%	

Source: (DWAF 2004)

[a] Includes the component of the Reserve for basic human needs at 25 l/person/day

[b] Mining and bulk industrial that are not part of urban systems

[c] Includes water for thermal power generation only, since water for hydropower, which represents a small portion of power generation in South Africa, is generally also available for other uses. (For ease of direct comparison with Eskom these numbers have not been adjusted for assurance of supply; the quantitative impact of which is not large)

[d] Quantities given refer to impact on yield only. The incremental water use in excess of that of natural vegetation is estimated at 1,460 million m^3/a

Table 2.4 Reconciliation of the requirements for and availability of water for year 2000 (million m³/a)

	Water management area	Reliable local yield	Transfers in[a]	Local requirements	Transfers out[a]	Balance[b, c]
1	Limpopo	281	18	322	0	(23)
2	Luvuvhu/Letaba	310	0	333	13	(36)
3	Crocodile West and Marico	716	519	1,184	10	41
4	Olifants	609	172	967	8	(194)
5	Inkomati	897	0	844	311	(258)
6	Usutu to Mhlatuze	1,110	40	717	114	319
7	Thukela	737	0	334	506	(103)
8	Upper Vaal	1,130	1,311	1,045	1,379	17
9	Middle Vaal	50	829	369	502	8
10	Lower Vaal	126	548	643	0	31
11	Mvoti to Umzimkulu	523	34	798	0	(241)
12	Mzimvubu to Keiskamma	854	0	374	0	480
13	Upper Orange	4,447	2	968	3,149	332
14	Lower Orange	(962)	2,035	1,028	54	(9)
15	Fish to Tsitsikamma	418	575	898	0	95
16	Gouritz	275	0	337	1	(63)
17	Olifants/Doring	335	3	373	0	(35)
18	Breede	866	1	633	196	38
19	Berg	505	194	704	0	(5)
	Total for country	13,227	0	12,871	170	186

Source: (DWAF 2004)

[a] Transfers into and out of water management areas may include transfers between water management areas as well as to or from neighbouring countries. Yields transferred from one water management area to another may also not be numerically the same in the source and recipient water management area. For this reason, the addition of transfers into and out of water management areas does not necessarily correspond to the country total. The transfer of water from Lesotho to South Africa is reflected in the tables as being from the Upper Orange water management area.

[b] Brackets around numbers indicate a negative balance.

[c] Surpluses in the Vaal and Orange water management areas are shown in the most upstream water management area where they become available (that is, the Upper Vaal and Upper Orange water management areas).

in deficit (with requirements larger than the available water). For four of the remaining nine, the surplus was less than 10% of the available yield.

However, these balances must be understood in the context of both the hydrological conditions of the country as discussed earlier in this chapter, as well as the change from one water act to another in 1998.

It was explained earlier that very little of the surface runoff is usable without the creation of storage in the system. Dams are built as water requirements grow. This creates a temporary surplus until the requirements has increased beyond the available water, after which another dam is built. It would thus invariably mean that with good planning, at no stage in time, large surpluses would exist.

The existence of deficits could probably be explained from three main angles. The National Water Act (NWA) (RSA 1998) introduced the concept of the Reserve and the water required to sustain the ecological Reserve is in some cases adding a considerable additional requirement on the systems. This was obviously taken into consideration in the calculation of the balances shown in these tables, while it was not taken into account under the previous water act. This effectively means that the ecological requirements are not fully supplied at this stage in many catchments. One of the big challenges of implementing the NWA is to make the water available for the ecological Reserve without causing economic and social disruption for the existing lawful users. Another reason for the deficits is again the complex hydrology of the country. When the large water resource developments started in the early part of the twentieth century the hydrological records were almost non-existent and in many cases even today remains deficient. The yield and subsequent allocations from a resource is done on the best available information at the time, but it is often found that the addition of more information as time progresses reflects more severe droughts, which then decreases available yield. Shortages like this can only be solved by adding more storage to the system, using water more efficiently and reducing the use from the resource, or by cutting the allocations with a resulting decrease in production.

To this already complex picture must still be added the issue of redressing the inequities of the past caused by the policies that excluded the black population from adequate access to the water resources.

It is possible to increase the availability of the resource by building more dams and in many cases this will have to be done. The NWRS shows that there is the potential to add major dams of a total of 5,410–32,410 million m^3 storage. The comparison of the existing and potential dams already tells a story. The majority of the dams in the country have already been built and the era of major dam building is history and cannot be repeated. Furthermore, in the majority of the catchments, the most economical dam sites have already been utilised for the existing dams. New dams will be very expensive to build and will yield water at much higher unit costs than before.

2.5 Main Challenges Facing Water Resources Management in the Country

It is clear that South Africa is water scarce and hence water will in future have to be used more and more efficiently. This applies to all sectors and implies that water will have to flow from least efficient sectors to those that will make better use of it, i.e. from low to high value productive economic activities.

New water is going to become very expensive and will invariably mean that sectors of economic activity will have to pay more for water and the government will have to contribute large subsidies for social developments.

More water to increase land under irrigation will be very limited. This is not a very popular message especially where some of the provinces in the drier parts of the country still believe their development route goes through irrigated agriculture. There are a few areas in the country where the competition for water will be limited and irrigation will for a long time to come be the sole user of the water. However the cost of building dams to supply this water will in many cases be high and government will have to heavily subsidise such development as was done in the past. However, the subsidies will relatively be much larger than for the early schemes, when the cheap dam sites were still available.

Water quality presents a major challenge downstream of all urban areas. Compliance of the towns with sewage standards is currently a major problem and this will have to be addressed urgently.

Coastal cities still discharge all or major volumes of effluent into the oceans. This represents a potentially huge resource that will most likely be cheaper to utilise than the desalination of seawater. This option will have to get better attention.

Supplying the ecological Reserve in many cases would mean the reduction in water availability for current and future off-stream use.

Redressing inequities will mainly have to be done by reallocation of existing water (and land).

International arrangements with neighbouring countries on shared rivers are expected to become a matter of key concern and tension as these countries' developmental needs for more water expand.

References

DWA (Department of Water Affairs) (1956) Water Act. Republic of South Africa, Pretoria

DWA (Department of Water Affairs) (1986) Management of the water resources of the Republic of South Africa. Pretoria, Department of Water Affairs

DWAF (Department of Water Affairs and Forestry) (1997) Overview of the water resources availability and utilisation in South Africa. Department of Water Affairs and Forestry, Pretoria

DWAF (Department of Water Affairs and Forestry) (2004) The national water resource strategy, 1st edn. Department of Water Affairs and Forestry, Pretoria

RSA (1998) National water act. Government gazette, vol 398. Office of the President, Cape Town, 26 August 1998

SANCOLD (South Africa National Committee on Large Dams), (1994) Large dams and water systems in South Africa. SANCOLD, Pretoria

Chapter 3
Water Services in South Africa 1994–2009

Kathy Eales

Abstract Since 1994, South Africa's government has demonstrated profound commitment to achieving universal access to water and sanitation – through ambitious policy reforms, institutional restructuring and substantial funding for infrastructure development and operating subsidies. It has met its 2015 MDG goals for both water services and sanitation. But the country as a whole has underestimated what it takes to sustain effective water services: expanding access is comparatively straightforward when compared to the challenges of providing an ongoing service, with all the inherent challenges of strategic management, operation and maintenance, revenue administration and relationship building. Concurrent local government restructuring has meant that infrastructure expansion has generally outpaced the institutional development needed to run the new services. Moreover, top-down, target driven delivery has largely marginalised and alienated ordinary people. Many municipalities have not resourced their water services functions appropriately, and there are severe municipal skills gaps. In response to poor services, unresponsive councillors and corruption, social protest action soared in 2009. Government is now giving priority to a comprehensive turnaround strategy to strengthen local government and its service delivery performance. But the more services are expanded and upgraded, the greater the need for operating subsidies, because the levels of service being provided – certainly in urban areas – are not affordable to a large percentage of the population. Growing dependence of municipalities on grants means that far greater attention must also be given to the efficiency of providing services, to contain costs. Meanwhile leaks and losses associated with neglected maintenance create supply bottlenecks, and raise the costs of provision. Improved management of water services is imperative to enable the inclusive growth and development that South Africans need to escape from high levels of poverty.

K. Eales (✉)
Counterpoint Development cc, Pretocia, South Africa
e-mail: kea@iafrica.com

B. Schreiner and R. Hassan (eds.), *Transforming Water Management in South Africa: Designing and Implementing a New Policy Framework*, Global Issues in Water Policy 2,
DOI 10.1007/978-90-481-9367-7_3,

Keywords Water services • Poverty • Financial sustainability • Free basic water • Municipal finance • Infrastructure • Service delivery • Maintenance • Subsidies • Bill of Rights • Equity • Tariffs

3.1 Introduction

South Africa has much to be proud of in its water services sector: Constitutional rights to dignity, access to sufficient water, and a healthy environment; a concentrated focus on achieving universal access to water services, supported by vast capital grants and operating subsidies; provision of free basic water to ensure that price is not a barrier to access; an innovative incentive-based approach to regulating water and wastewater quality; and so on. High-level commitment, backed by substantial spending on infrastructure development and servicing subsidies, has led to some extraordinary achievements. Government figures indicate that by 2009, water services infrastructure had been extended to 96% of the population, up from 59% in 1994 (DWAF 2009a). Put differently, the number of additional people reached by new water services infrastructure since 1994 is far greater than the total population of Angola, Chile or the Netherlands.

By any measure, this is a remarkable achievement. But, in its commitment to meeting service delivery targets far more ambitious than the MDGs, government has underestimated the challenges of ensuring sustainable, ongoing water services. Providing a reliable ongoing supply of safe drinking water 24/7 is far harder than building the infrastructure needed to enable access. In many areas, problems in operating and maintaining infrastructure are leading to service breakdowns, and delivery short of expectations is fuelling increasingly vocal public protest.

The integrated nature of the water services process – across water resource management, environmental management, and technical, financial and administrative dimensions, with a necessary emphasis on managing the relationship between the service provider and many thousands of service users – makes it much more challenging than other sectors which combine infrastructure and servicing (Muller 2009). The stakes are higher too. Mismanaged water and sanitation services have enormous potential to cause severe hardship, illness and environmental damage. Mismanagement impacts most harshly on the poorest, but also on business, industry and tourism and their ability to provide jobs and generate tax revenue.

Good management of municipal water services is vital for national water security, particularly as climate change and unpredictable rainfall introduces new risks and uncertainties. This requires water services institutions that are competent to deliver reliable supplies of safe drinking water in a context of deteriorating raw water quality, rising demand and growing scarcity, and can manage the collection and safe treatment of wastewater without imposing harsh burdens on future generations. Far greater reliance on re-use and recycling is likely. These issues underline the importance of cost-effectiveness and affordability in water services, nationally, to safeguard the rights of access of the most needy and vulnerable citizens.

But there is growing acknowledgement of the gulf between the state of water services management, and the level of professional skills required to manage growing complexity.

The tension between infrastructure provision and on-going service provision capability has played out in a context of concurrent reform in the water sector and the local government sphere. Rapid decentralization of responsibility for the provision of water and sanitation, with massive spending and development to achieve universal service access swiftly, occurred in tandem with massive upheavals in the form and function of local government. Since 1994, the role of local government has been fundamentally reconceived as the primary driver and enabler of development, as well as the primary implementer of programmes to eradicate apartheid-based service disparities in water, sanitation and a range of other areas. Immense restructuring and transformation in local government has been necessary to lay the basis for this new role. One consequence is that national policy objectives do not yet align with the municipal capability to implement those policies. Another is that most municipalities have opted for a technocratic top-down approach to delivery, with decisions implemented with little public involvement. But many municipalities lack the skills needed to succeed with a technocratic approach; in too many instances, the result has been non-delivery, or development of poor quality infrastructure without the resources to keep it working.

The growing number of service protests – far more between January and August 2009 than in any full year since 1994 (CGTA 2009d) – are not simply a demand for 'better' services, delivered more effectively and efficiently. More importantly, they reflect a demand for a more inclusive mode of development (Pithouse 2009), where priorities and projects are shaped through engagement and confrontative dialogue (Galvin 2009), rather than by external delivery targets. This is likely to entail a more protracted approach to service improvement, driven less by patronage than through partnership; but without this, sustainable services that meet people's needs affordably are unlikely.

This chapter reviews some themes in water services delivery over the past 15 years, flags some lessons learnt the hard way and underscores the importance of more professional water services management, particularly as evidence of greater climate variability mounts.

3.2 Poverty and Inequality: the Context of Policy Formulation

Poverty, inequality and vulnerability define South Africa to its core, and provide the essential context for the path the country has pursued for water services. South Africa has the world's 33rd largest GDP, and is a middle-income emerging economy. Its population is just less than 50 million people. Average GDP per capita (purchasing power parity) was US$9,757 in 2007, but this figure obscures enormous disparities in income (UNDP 2009). According to UNDP figures, 43% of South Africans lived on less than US$2 per day in 2007 (Ibid.), as a result of structural poverty and chronic unemployment, compounded by the impacts of one in six adults being infected with HIV. The impacts of AIDS are rolling back decades of development: average life expectancy in South Africa has now fallen from 63 years in 1990 to 50.5 years in 2009, similar to what it was 40 years ago (ASSA, cited in Presidency 2009; UNICEF 2009). South Africa now has the biggest anti-retroviral programme in the world, but about 1,000 people die every day from AIDS-related problems, and there are an estimated 1,500 new infections each day (MRC 2009).

South Africa has one of the highest Gini co-efficients in the world, denoting extreme income inequality. The Presidency acknowledges a Gini co-efficient of 0.666 (Presidency 2009), with other sources rating it even higher (Bhorat et al 2009). The black middle class is growing rapidly, but the median income of the white population, which comprises about 9.2% of the total, remained more than ten times that of the black population in 2008 (Presidency 2009).

South Africa's poorest households are female-headed, and rural. 'Rural' is a very inadequate description of the wide range of settlement types clustered under this loose category by census enumerators, and suggests an easy distinction between urban and rural areas. The reality is more a continuum of settlements types and densities, with a range of different defining features. In the context of this discussion, rural areas include commercial farmland, where the majority of farm dwellers live in conditions of acute poverty and marginalisation. In the South African context, however, rural areas are more widely associated with the former 'Bantustans' or 'homelands', and include areas of subsistence farming and extensive settlements with no discernable economic base beyond welfare grants. Only a quarter of working-age adults in these areas are employed, compared to almost half in the rest of the country, and employment in the former Bantustans is typically at very low wages, with over half of those employed earning under US$135 a month in income in 2006 (Philip and Hassan 2008).

Although one third of households lived in rural areas in 2005, well over half – 57.1% – of all *poor* households live in rural areas (Armstrong et al 2008) and the incidence of poverty in rural areas is more than double the corresponding rate for urban areas. Table 3.1 highlights the severity of rural poverty, and its gendered dimensions.

In 1994, four in ten people lacked access to a basic supply of safe drinking water, and more than half had inadequate sanitation (RSA 1994), with backlogs most severe in rural areas. Urgent interventions were required to address gross inequality and underdevelopment across virtually every dimension of the state, and the new government quickly embarked on a systematic overhaul of the structure, objectives and statutes of government.

Table 3.1 Income distribution in South Africa, per quintile (Armstrong et al. 2008), sourced from the 2005/06 Statistics South Africa *Income and Expenditure Survey*

	Quintile 1	Quintile 2	Quintile 3	Quintile 4	Quintile 5	All
Rural households	63	46.8	34.5	20.4	9.8	34.9
Urban households	37	53.2	65.6	79.6	90.2	75.1
Female-headed h'holds	54.6	48.7	39.1	28.7	23.1	38.9

Two core documents, each arising from intensive national consultation and negotiation processes, shaped the new government's course of action. The first was the 1993 Interim Constitution, which spelled out the institutional framework of the new South Africa. The second was the 1994 Reconstruction and Development Programme (RDP), which effectively constituted the ANC government's political manifesto. The RDP stressed that,

'The national government wishes to unlock the political and creative energies of the people and bring government closer to the people … for the first time in South Africa's history, emerging democratic local authorities must work with community-based organizations and NGOs to establish minimum conditions of good governance and to implement effective development projects' (ANC 1994)

The fundamental premise of the RDP was that people who are affected by decisions should take part in making them, and that development would occur through the growing empowerment and self-reliance of communities in shaping their own destinies. Second on its list of six basic principles was the assertion that, 'Development is not about the delivery of goods to a passive citizenry. It is about active involvement and growing empowerment' (ANC 1994). The themes of participatory democracy and people-centred development cross-threaded the RDP's five key programmes, notably those of meeting basic needs; developing human resources; democratizing the State and society; building the economy; and implementing the RDP. Water was an essential ingredient in each of these programmes.

The Constitution, adopted in 2006, added a right of access to water to the Bill of Rights. Local government was given responsibility for ensuring the provision of services, while national and provincial government were mandated to strengthen the capacity of municipalities to perform their functions, and to regulate their performance.

3.3 The Policy Framework for Basic Water Services Provision

In November 1994, DWAF published a White Paper on Water Supply and Sanitation, which gave effect to the RDP in the water sector. The policy's goal was 'to ensure that all South Africans have access to essential basic water supply and sanitation services at a cost which is affordable both to the household and to the country as a whole' (RSA 1994).

The White Paper's foremost concern was equity for the poorest. It acknowledged the diversity of water supply arrangements countrywide, but its focus was on enabling access to a basic water supply in the country's neglected rural areas through community water and sanitation initiatives.

Ahead of any international benchmarks, the White Paper formalised the concept of a basic water supply as the minimum standard, and quantified this as 25 l per person per day within 200 m. This was the minimum required for direct consumption, food preparation and personal hygiene, in line with government's commitment to maximise coverage and ensure all South Africans had access to basic water and sanitation within 7 years. In line with its principle of 'some for all, not all for some', the White Paper recognised that the amount should be adjusted over time.

The White Paper drew heavily on international experience, particularly the failure of technicist supply-driven approaches to infrastructure delivery. Echoing the RDP, its core theme was that development should be people-driven, and that 'the provision of services in poor communities will fail if the people themselves are not directly involved. The involvement and empowerment of people is thus a cornerstone of the approaches proposed.' A complementary strand was the principle of payment for services:

> If the community expects some outside agency to be responsible for keeping their supplies going, they will have no control over the processes and lose leverage and ownership. Responsibility for keeping the service going is placed with a remote authority and accountability is lost.

The White Paper argued that services should be provided and paid for in a manner which would not require on-going government funds to keep them running, and should be self-financing. The only exception would be where poor communities were unable to afford basic services: 'Government may subsidise the cost of construction of basic minimum services but not the operating, maintenance or replacement costs.' A social tariff, covering only the operating expenses, would be charged for the minimum level of service, which was a communal water source. For higher levels of service, the full cost of supply would apply.

The White Paper endorsed a sliding tariff scale, with three tiers. The first was a life-line or social tariff, to cover basic human need, for a quantity not exceeding 25 l per person per day. The second was a normal tariff, for amounts up to 250 l per person per day, provided at cost (operation and maintenance plus capital), including a portion to recoup the loss incurred from the life-line tariff. The third tier was a marginal tariff for consumption of water above 250 l, assumed to be for luxury uses, and consequently, charged at the marginal cost, which was defined as the present day cost of the latest or next augmentation scheme. The principle of a rising block tariff, designed to cross-subsidise the poorest users, was thus a central feature of post-1994 sector policy.

The 1997 Water Services Act gave legislative effect to the 1994 White Paper and the constitutional right of access to water, and wrote the concept of a basic level of service into law (Muller 2007). A core principle of the Act was the distinction, at local level, between a municipal Water Services Authority, and the Water Services Provider. Responsibility for ensuring access to, at least, basic water and sanitation services was assigned to those municipalities designated as Water Services Authorities. Actual provision was the responsibility of designated water services providers, which could be the technical section of the same municipality, another municipality, or an external contracted agency. The key distinction was between the authority function – responsible for setting policy, ensuring universal access to at least a basic service and regulating service provision to ensure effectiveness, efficiency and sustainability – and the provider function, which was responsible for executing the policies and strategies of the authority and delivering the actual service.

When performing the functions of provider itself, a water services authority was required to distinguish between its authority and provider functions, and account separately for them.

However, subsequent developments would show that this elegant, conceptual separation between local regulator and provider was seldom achieved, primarily because the local government sphere was preoccupied with far-reaching transformation processes of its own, with a different agenda.

3.4 Expanding Access to Water Services

The immediate challenge facing the new government in 1994 was to assemble a single public administration. Apartheid South Africa was comprised of eleven separate administrative and political territories: ten so-called 'homelands' scattered over 50 separate land portions, and the dominant Republic of South Africa, governed by a tri-cameral parliament for three minority racial groups.

In the towns and cities of so-called 'white South Africa', separate municipal service delivery administrations addressed the needs of each of the four official racial groups, with services in black townships generally inferior and often rundown. Rural areas were served either by provincial administrations or regional services councils at district level.

The ten 'homelands' were run by large bureaucracies, virtually as an end themselves, with poor productivity and corruption well-entrenched. The legal sophistry of apartheid had created a vast unaccountable muddle of delegated jurisdictions and overlapping administrations to postpone the inevitability of democratic government, and effective service delivery to all South Africans was simply not on the agenda pre-1994.

As the 1990s progressed, 2000 became the deadline for putting in place a framework of municipalities that would permit wall-to-wall local government countrywide. Massive restructuring would be necessary to integrate and reconfigure the administration of the country into a form able to tackle the enormous development challenges it faced. Yet, despite the structural integration and change in leadership that followed, the staffing and ethos of the previous administrations formed the backbone and shaped the practise of the new administration, with a remarkable degree continuity (McLennan and Munslow 2009).

In 1994, South Africa's urban landscape was already in flux, with the task of integrating local race-based administrations already starting to get underway. Administrations established to run 'white areas' were having to come to terms with the fact that they were no longer in charge, and that the majority of resources could no longer go to serving only the needs of a privileged minority. There were, however, at least core functional administrations in place to plan and manage the steady extension and upgrading of water service coverage, even if their spatial boundaries were extended hugely.

In the rural areas of the former 'homelands' where backlogs were most acute, there were no municipal structures to effect delivery, just the vastly fragmented remains of the apartheid state architecture.

The new ANC government assigned the new Department of Water Affairs and Forestry (DWAF) responsibility for tackling water services backlogs in rural areas, until local government structures were in place in all areas. Pre-1994, DWAF had been concerned, primarily, with water resource management, and ambitious infrastructure schemes to serve the country's macro-economy. Second-tier management, routing bulk supplies from dams to mines, industry, municipalities and other centres of demand, was essentially the domain of the water boards, such as Rand Water and Umgeni Water. In the former 'homeland' areas, third tier management – retail water services and domestic water provision – was now added to DWAF's mandate. Water services in towns and cities remained the responsibility of the municipal administrations that ran them (Schmitz 1999).

DWAF was able to hit the ground running in 1994 in its expanded role with well-formulated policies and a practical implementation strategy, thanks to the far-reaching preparatory work of an extra-parliamentary forum for water-related matters established in 1992, called the Standing Committee on Water Supply and Sanitation or SCOWSAS. The imprint of SCOWSAS is evident in the RDP and its recognition of the importance of access to water as an enabler of development (Muller 2007). Equally, DWAF had benefited enormously from its participation in a broad Drought Forum of political parties, government and non-governmental agencies, which sought to ensure that previously neglected rural communities benefited from government and donor-funded drought relief initiatives in the early 1990s. The Drought Forum's approach was community-based and focused on setting up and supporting CBOs to manage and operate local water schemes. It was, subsequently, widely adopted in the early phases of DWAF's Community Water Supply and Sanitation (CWSS) programme (Galvin and Habib 2003).

Within DWAF, a new Community Water Supply and Sanitation (CWSS) division was established in 1995, and the department's budget ratcheted up swiftly to fund implementation of rural water and sanitation projects. DWAF rapidly assembled an impressive delivery machinery, able to extend coverage to areas where previously there were none. Projects were selected, based on a nationally co-ordinated planning process, and no ceilings were applied to per capita subsidy amounts for water. Recipient communities were expected to pay for the operating and maintenance costs of installations, but no contribution was required towards capital costs. Responsibility for overseeing the operation and maintenance of the new schemes would, in time, be transferred from DWAF to municipalities (Savage and Timm 2003, unpublished report).

Water boards and water sector NGOs were enlisted to assist delivery. There was a strong commitment to community participation, and in the case of the Mvula Trust and its partners, to demand responsive approaches and community management. The Mvula Trust and its partners were particularly effective in working on the ground with CBOs, equipping them to plan, manage and, subsequently, operate more than 300 water and sanitation projects. Many were remarkably effective and successful (GTZ 1998). A handful of these projects survived for over a decade, still run by community structures separate from local government and funded by residents. Possibly the last remaining community-run project, in Nhlungwane in southern KwaZulu-Natal, was absorbed into municipal servicing in early 2008.

Capacity to implement demand-responsive projects was limited, however, and the realities of mobilising poor communities to establish village-level structures capable of co-planning and managing projects and collecting their designated contribution of 10% of the capital cost (a feature of the Mvula approach), constrained the pace of progress. Participatory development has its own time frames, and cannot easily be reconciled with the requirements of external delivery schedules. Participatory approaches increasingly came to be associated with slow delivery, notwithstanding their other merits.

Further capacity was brought in from 1997, with the launch of the Build Operate Train and Transfer (BOTT) programme in the four provinces with the greatest backlogs. Consortia of private sector firms, at times with NGO involvement, would deliver a package of infrastructure, accompanied by community consultation and training. Project performance criteria specified the use of local labour, and balanced representation of men and women on project steering committees.

The enormous pressure to achieve tangible results, and the scale of resources being allocated to rural water and sanitation projects, steadily tipped the balance towards short-cuts in establishing village water committees able to manage and maintain scheme. In many instances, these short-cuts compromised their ability to function effectively – leading community-level planning processes, overseeing the recruitment and training of labourers, collecting contributions from local residents, and ultimately running basic water supply schemes. But voluntary community management structures were not necessarily appropriate in many of the more complex schemes. Where a steady water supply required a committee simply to maintain a pipeline from a safe mountain spring, the risks of failure were minimal, but more elaborate schemes, requiring diesel or electricity purchases, maintenance and regular payments by users, required greater training and commitment (Muller 2007). The responsibilities of users were often not adequately communicated to them, leading to hostility and resentment when they were expected to self-fund malfunctioning schemes. Where local community structures were unable or unwilling to keep schemes functioning, water supplies tended to fail.

An important part of DWAF's response to evidence of weak schemes and poor cost-recovery was to re-assert the importance of involving residents in the planning and management of schemes, and to refine its approaches to building effective local structures able to oversee local project implementation and the subsequent operation and maintenance of village water schemes. In some regions, however, where local government restructuring was gaining momentum and municipalities were establishing their own structures and planning forums, these community water and sanitation committees resembled a parallel rural local government system, prompting tensions between sectoral and municipal-led initiatives (Savage and Timm 2003; DPLG 2001). Moreover, some municipalities maintained that DWAF was high-handed in its approach, by building infrastructure and expecting them to take over ownership and maintenance without having adequately engaged them in planning or budgeting (DPLG 2001). This was particularly true from 2000, following South Africa's second democratic elections and the establishment of representative local government structures country-wide.

3.5 Local Government Restructuring

In its magnitude and complexity, the process of local government transformation dwarfs any other institutional change in the history of South Africa (DPLG 2001). Since 1994, local government transformation has aimed at making municipalities more accountable, financially sustainable and able to deliver critical services. Municipal structures were reconfigured or created to include areas not previously served by local government, and to deliver infrastructure in a manner that would stimulate economic growth and reduce inequality (Cartwright 2003).

It is difficult to overstate the scale of the changes that municipalities underwent in the 1990s and beyond. These included the integration of racially-distinct municipal administrations, new structures for democratic and accountable government, new systems for improved operational and financial management, and an entirely new conception of the role of local government. Municipal boundaries were redrawn twice. South Africa's new Constitution significantly altered the status, powers and functions of local government, and redefined it as an autonomous sphere of government that had developmental duties that extended beyond its core service functions and included ensuring the economic and social welfare of its constituents. The 1998 White Paper on Local Government detailed this vision of 'developmental local government' as, 'working with citizens and groups within the community to find sustainable ways to meet their social, economic and material needs and improve the quality of their lives' (RSA 1998). This was followed by a systematic overhaul of legislation. By late 2000, 843 pre-existing municipalities had been consolidated down to 284, and classified into one of three categories – six became single-tier metropolitan municipalities; 232 became local municipalities, and 46 became district municipalities, made up of between five and eight local municipalities each.

In the former homeland areas, entirely new municipalities were created. They faced enormous pressure to deliver on their new mandate, as well as immense challenges in providing functional, viable services in a context of chronic poverty and under-development.

New boundaries led to major staff movements as personnel were transferred between newly-defined municipalities. Many posts were rationalised as existing municipalities amalgamated, and a moratorium was placed on the recruitment of new staff to key positions. The impacts of rationalising technical and engineering positions were particularly severe. The scale of these changes put great strain on structures that were already under-resourced, and there was uncertainty over jurisdiction, budgets and basic planning data. In parallel, water and sanitation infrastructure programmes were being run locally by different sector departments and agencies – DWAF, water boards and NGOs in rural areas, as well as the Department of Provincial and Local Government and municipal programmes in peri-urban and urban areas – with different messages and policies, different default service levels and different subsidies and grants. Municipal leaders, at times, had to intervene to explain to their constituents why some settlements, supported by different government programmes, were apparently getting preferential treatment, while in others residents had to contribute their time, materials or cash to the project roll-out. There was immense confusion and discontinuity as new systems began to take over from the old, and schemes and services began to be transferred from DWAF to local government.

In 2000, all district municipalities were declared Water Services Authorities, and many had to gear up from nowhere to develop policies and programmes to drive service delivery. Just 3 years later, in mid-2003, these authorisations were reversed in some areas when national government finalised the allocation of municipal powers and functions for water and other services, and designated 183 municipalities as Water Services Authorities. In some instances, municipalities which had been playing the role of both authority and provider, now had to hand over one or both of these responsibilities to another municipality, along with the staffing and resources that went with these functions. There were inevitable lags and lapses in the transfer of funds from national government – funds for capital expenditure might be assigned to the district municipality, while funding transfers for the recurrent costs of provision went to the local municipality. Equally, the infrastructure and equipment needed for service delivery may have been be registered as an asset of one municipality, while the responsibility for providing the service now lay with an entirely different structure. This led to confusion over where the responsibility for funding maintenance lay, and resulted in significant gaps in asset management and budgeting. In many municipalities, essential maintenance of infrastructure needed to support effective service delivery was simply neglected.

3.6 Fiscal and Financial Restructuring

The National Treasury, meanwhile, was steadily transforming the national fiscal and financial policy framework. One core objective was to increase the flow of funds through direct and indirect transfers to local government. Another objective was to strengthen municipal financial management and accountability systems, with the 2003 Municipal Finance Management Act elaborating on a range of mandatory systems and procedures to promote good governance (Savage 2007).

Two new local government funding mechanisms have particular relevance for water services provision – the Equitable Share, and the consolidated Municipal Infrastructure Grant (MIG). The Equitable Share was introduced in 1998 to give effect to the Constitutional requirement that provincial and local government should have an equitable share of national revenue, commensurate with their functions.

It had three components, intended to support municipalities' *recurrent* costs – a basic services grant, to subsidise basic services for the poor; an institutional capacity-building grant to support individual municipalities that were unable to raise enough revenue to establish their core administrative infrastructure; and an intergovernmental grant allocation, based on a proportion of grants paid in previous years. From 2001, largely in response to growing evidence that poor households were struggling to afford the cost of the new services being provided to them, the Equitable Share was significantly increased to accommodate the cost of the new Free Basic Water Policy of providing 6 free kl of potable water monthly to every poor household. The policy was announced in September 2000 to ensure that no poor person should be denied access to a basic supply of water based on affordability. This represented a significant shift in government policy, with government now co-funding a portion of the operating costs of water services, as well as electricity, waste removal and sewage.

The MIG, conversely, provided *capital* funding to enable municipalities to invest in the infrastructure needed to provide basic services to the poor. Introduced from 2004, MIG consolidated several different national funding streams into a comprehensive grant, with the amount available per municipality over the next 3 years, published each year. For the first time, municipalities could plan multi-year programmes with reasonable certainty regarding their grant funding allocations for subsequent years. MIG was based on a formula that took into account the number of people still requiring basic infrastructure. Over half of the annual allocation was earmarked for new water and sanitation infrastructure. The funds were allocated per municipality and could be accessed directly from the Department of Provincial and Local Government, rather than via the sectoral programmes of the different line departments.

The introduction of MIG also meant that from 2004, municipalities accounted to DPLG, not DWAF, in how they spent capital grants for infrastructure in the water services sector. DWAF's role in relation to capital funding was limited to setting design standards and scrutinizing project technical reports during the preliminary design phase. This diminished DWAF's influence over how projects were conceived, implemented and monitored. As municipalities took up their new mandate, with increased funding and autonomy, project planning and implementation became increasingly technicist, with less and less emphasis on the social capital and maintenance dimensions of sustainable services.

3.7 The Diversity of Local Government

It is important to understand the immense diversity in the form and capacity of the 283 municipalities nationally. There are three broad types of municipality: metropolitan, district and local. Metros have the most resources and the greatest degree of autonomy. Elsewhere there is a two-tier system of local government, with five to eight local municipalities (LMs) clustered under a district municipality, with widely ranging

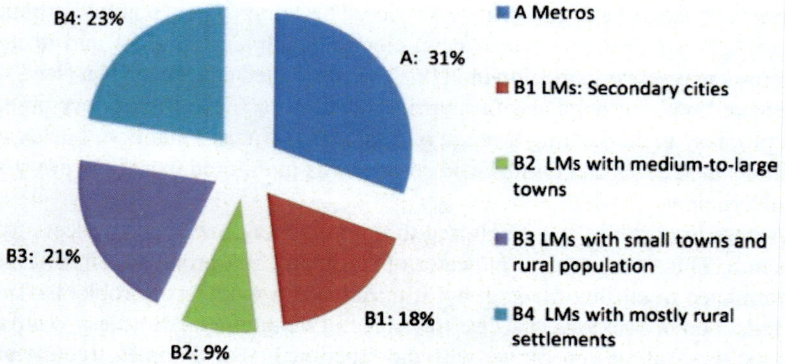

Fig. 3.1 Types of municipalities

powers and functions. In some areas, the district municipality is the dominant authority, with primary responsibility for provision of services; in this case it is the primary recipient of transfers from government. Elsewhere, district municipality powers and functions are more limited, and its role is primarily co-ordination; this frequently leads to contestation around decision-making authority and allocation of resources.

National Treasury classifies municipalities nationally into seven categories, based on population, percentage of urban population, size of budgets and spatial context; this is useful for understanding the different challenges they face. Figure 3.1 shows the distribution of the different forms of municipalities, and the percentage of households nationally they serve.

Poverty and service deficiencies are most acute in B4 municipalities in the former 'homeland' areas, with the most financially and administratively stressed municipalities falling in the B3 category (CGTA 2009e).

3.8 Targets and Backlogs

In the 1994 White Paper on Community Water Supply and Sanitation, government had promised universal access to basic services within 7 years. The scope and scale of the task was daunting, but there was no debate about government's commitment to overcoming the inequities of the past. Since 1994, government has consistently tracked progress in extending access, and has committed vast resources to enable the achievement of universal coverage; alongside this, it has put immense pressure on the different agencies of government to extend services as swiftly as possible.

The 2003 Strategic Framework for Water Services – essentially an updated and expanded White Paper for water services, which reflected municipalities' leading role in water services provision and DWAF's changing role – included a range of new delivery targets. The two most important related to water and sanitation coverage: by 2008 for water, and 2010 for sanitation, all people in South Africa were to have access to at least a basic functioning service. This commitment featured prominently in the ANC's national and provincial election campaign in 2004 and in its local government election campaign in 2006. The Mbeki-led administration (1999–2009) is credited with an emphasis on managerialism to achieve its delivery objectives, and progress in eradicating a range of backlogs – water, sanitation, bucket toilets, facilities in schools and clinics, and so on – was monitored closely in every sphere of government.

Access to services was monitored through tracking backlogs in access to infrastructure. This was a critical indicator of inequality and provided a useful index of performance in closing the gap; but it introduced a number of problems. Tracking the reduction of backlogs in access to water and sanitation made access to infrastructure the over-riding imperative, with the functionality of the infrastructure and the quality of the service provided secondary. Secondly, it focused spending on new capital development, frequently leading to neglect of maintenance and rehabilitation of existing infrastructure and, increasingly, of the new infrastructure as well. Thirdly, it diverted attention away from the long-term operating costs and requirements of different infrastructure options and service packages. With access to taps and toilets being provided at no cost to the beneficiaries, and municipalities receiving funds from national government for both capital and operating costs, the life-cycle costs of different service packages were seldom assessed rigorously – the priority nationally was to expand access to services as swiftly as possible, utilising funding made available by national government on a 'use it or lose it' annual basis.

The 1994 White Paper on Community Water Supply and Sanitation aimed for breadth of coverage, not depth, using the principal of 'some for all' not 'all for some': the intention was to achieve universal coverage with a basic level of service as the first priority, before providing higher service levels. But the Community Water Supply and Sanitation White Paper, and the White Paper on Basic Household Sanitation (drafted and influential from 1996, although it was only finalised in 2001) were perceived widely as DWAF policies that were applicable only in the rural areas served and supported by DWAF. Outside of dispersed rural settlements, most politicians and officials aspired to deliver significantly more than a basic level of service, irrespective of cost. The 2003 Strategic Framework for Water Services – developed largely by DWAF – re-affirmed the concept of breadth not depth as the first priority, and introduced the notion of a water ladder. Provision of at least a basic water service was the first step up the water ladder, but government was committed to progressively improving levels of service over time, moving on to intermediate levels of service as the next step – a yard tap, not a public standpipe, for example (RSA 2003).

In urban settlements, reticulated water and sewage networks were technically feasible, and it was hard to promote intermediate or differential service levels, like low-pressure water systems or yard taps, when white households in adjacent settlements had in-house taps and flush toilets. Moreover, consideration of the cost to poor households of different services packages was generally not a significant factor when planning new services. In many of the former homelands, services had been provided free, funded by the South African fiscus; and from the early 1980s, rent and service boycotts were used widely in South African's townships to cripple local administrations perceived as illegitimate, and non-payment remained deeply entrenched. In Soweto, for example, less than 12% of residents were paying their water services bills in 2003 (Johannesburg 2003). Municipal officials and politicians taking decisions on service levels evidently assumed that grant funding and transfers would make good any deficit.

The new municipalities faced enormous pressure from their constituents, provincial and national government and the ruling party to extend service coverage, and worked hard to deliver; they provided high level services as the default in urban settlements. With government funding huge housing and infrastructure development, beneficiaries understandably wanted an in-house tap, not a yard connection or stand-pipe, and a flush toilet, not a dry toilet. Where a lower level of service was offered, residents frequently brought the project to a halt until there was a commitment from local politicians and officials to provide a higher level of service, either through sourcing additional funds and changing the output of the current project, or proceeding with the current project, with a promise to upgrade it within a defined period. Provincial housing delivery programmes provided free houses to poor families, with house taps and flush toilets as the default; in many areas new settlement development went ahead without even confirming that the municipality had the water supply or bulk infrastructure to support a large new housing development with high-level services, let alone the revenue to fund subsidised service provision.

Service coverage expanded steadily across the country, as MIG funding steadily increased. The notion of 'backlogs' was, however, become increasingly problematic. Government had set firm targets – universal access to water by 2008 and to sanitation by 2010 – but municipalities were finding that it was not simply a matter of pouring resources until full service coverage was reached. The backlogs were not static. The country's population was growing rapidly, while average household size was falling; this meant that while the population grew by 20% between 1996 and 2008, the number of households grew by 39%, with profound implications for the number of households requiring service points.

In attempting to eradicate backlogs, government faced a moving target. Moreover, households themselves were moving. Migration dynamics are complex and highly varied, but the notion of urbanisation as an orderly one-way movement from rural to urban areas does not adequately characterise settlement dynamics in South Africa. The country has a long history of migrant labour, and even though the pass laws were scrapped in the 1980s, the extent of migrancy is now even greater than during apartheid. In the face of chronic unemployment and deep-rooted poverty, particularly in the former homelands, households pursue multiple livelihoods strategies to access income opportunities, with family members moving for varying periods to places perceived to offer better opportunities. Added to this, people come to towns to access better schools, medical care and housing, while others straddle a home in a town or city and a family base somewhere else where they might send their children or other family members for a range of reasons (Kok et al. 2005). One consequence is that many families need more than one service point. Moreover, many households move as work opportunities come and go;

municipalities complained that while they might have eradicated all backlogs 2 years previously, the situation subsequently was entirely different.

Tracking backlogs became profoundly politicised in the second decade of ANC government, as senior politicians committed themselves to ensuring that targets would be met, and respective Water Affairs Ministers and senior officials exhorted municipal representatives to intensify their efforts to meet sectoral targets. Meeting the backlogs was increasingly becoming a numbers game, with less and less connection to access to good quality – let alone sustainable – services. DWAF, for example, compiles data which is quoted widely by the Presidency and elsewhere; backlog reduction is tracked against a 1996 baseline using project implementation data. Figures reported for the financial year ended March 2009 indicated that 96% of households had access to water supply infrastrucfture, up from 59% in 1994 (DWAF 2009a). In a footnote to a table of figures, DWAF sounded a caveat: the figures showed that 96% of households had access to the *enabling infrastructure*, but not necessarily to functioning services. Closer reading of the data reveals that the number meeting the RDP minimum standard was 91%.

However, this indicator excludes informal settlements, because improved water services for shack-dwellers is the responsibility of provincial Departments of Human Settlements through new housing developments or in-situ settlement upgrades. DWAF does not regard improved water services in informal settlements as its responsibility, and therefore excludes at least 10% of the population when reporting on access to water services nationally. Relatively few people are aware of the precision of DWAF's reporting indicator: it tells us that 86% of people (acknowledging the omission of people living in informal settlement statistics) have been reached by infrastructure development programmes. What a figure of 96% suggests is that the goal of universal coverage has virtually been achieved, allowing some room for complacency. This means it is the wrong indicator, because it fails to reflect the level of crisis growing in South Africa's water sector.

3.9 The Growing Crisis of Water Services

The nature of this crisis has many facets, and a flurry of government reports has quantified its dimensions.

3.9.1 Access to Basic Services

Access to a basic water services remains poor for those in the lowest quintile of the population by income. Just 41.6 % of households in the poorest 20% of the population by income level had access to piped water in 2006, according to the 2006 Stats SA General Household Survey (Statistics South Africa 2008). Notwithstanding immense effort and expenditure by government, not much has changed for South Africa's poorest and most vulnerable people since 1994. In 2008, six municipalities had water services backlogs above 70% of households, and 46 – mostly small municipalities with populations under 50,000 households – reported backlogs of over 40% (CGTA 2009c). Moreover, the distribution of water backlogs has barely changed in the past decade. The greatest backlogs are still in former 'homeland' rural areas in the three poorest provinces (KwaZulu-Natal, Limpopo and the Eastern Cape) and still twice as high as in any of the small towns or cities in the rest of the country – including major metropolitan areas where there has been significant growth in informal settlements. The bulk of the task of eradicating service backlogs continues to lie, primarily, with the municipalities least resourced to deal with the challenges.

As the head of the department responsible for Local Government told Parliament in June 2009, 'Service delivery to the poor remains our greatest challenge' (Africa 2009).

3.9.2 Quality of Service

Despite significant service improvements in some municipalities, the general trend is not positive. The percentage of households who rated the quality of water services provided by local municipalities as 'average' or 'poor' increased from 24% in 2005 to 39% in 2008 (Stats SA 2009b). Households who reported water interruptions at least once a month or more often increased from 19.3% in 2002 to 31.4% in 2008, with figures in the Free State Province trebling, from 16.7% in 2002 to 42.6% in 2008 (Stats SA 2009b); this may be related to the rapid increase in water-borne sanitation coverage associated with the bucket toilet eradication programme, in a context of widespread water shortages (See Chapter 4 for further discussion of the bucket eradication programme.)

Evidence of declining water quality in some areas is of particular concern, given the high vulnerability to water-borne diseases of people with compromised immune systems. Typhoid killed at least 13 people in Delmas, near Johannesburg, in mid-2005, and over a thousand cases of diarrhoea were reported 2 years later when problems with chlorination recurred (DWAF 2007b). In April 2008, over 140 babies and infants died in the uKhahlamba district of the north Eastern Cape from a combination of poor nutrition, faecal contamination of water and inadequate hospital treatment (DoH 2008). In the summer of 2008/2009, cholera spread from Zimbabwe into Limpopo and Mpumalanga provinces and beyond, causing the deaths of at least 51 people in South Africa. The rate of infection was exacerbated by water services failures in some Limpopo municipalities, through delivery of unsafe water, or no delivery at all. A senior official of the Department of Health told members of the Parliamentary Portfolio Committee on Water Affairs in February 2009 how problems in local formal water supply systems had worsened the epidemic: inadequate chlorination, poor maintenance of treatment works and pump stations, interrupted water supplies, provision of raw water through taps; where water supplies failed, local residents reverted to untreated water from boreholes and rivers, some of which were contaminated by poor municipal wastewater treatment or sewage spills (Benson 2009).

In early June 2009, eight villagers died in the Eastern Cape village of Mpheko, near Mthatha, after drinking contaminated river water when the municipal water reservoir ran dry. Improved services and new housing projects in the area promoted in-migration and rapid growth in many small settlements, far beyond the design capacity of the existing water supply infrastructure. Intermittent municipal water supplies were reported in Mpheko Village from December 2008, and many households reverted to using local springs. But severe drought by mid-2009 meant that most local groundwater sources were dry, leaving those without access to municipal water to draw from local rivers. The deaths followed an outbreak of diarrhoea and vomiting in later May 2009 (SAPA 2009; Groenewald 2009; Daily Dispatch 2009; Yeld 2009).

These problems are likely to worsen. A March 2009 *Water and Sanitation Audit Report,* commissioned by the Minister of Provincial and Local Government, found that 85% of the country's wastewater treatment works had a limited remaining useful life, while 90% of water treatment works were dilapidated (EU Municipal Outreach 2009); in April 2010, the Minister of Water Affairs formally requested R23-billion over the next 6 years from National Treasury, and noted that the full amount needed was close to double that (SAPA, 2010). However, at issue is not only the state of infrastructure, but the skills needed to run treatment works. A 2009 assessment found 114 municipalities grossly non-compliant with regulations stipulating the minimum qualifications levels needed for process controllers, plant supervisors and maintenance teams in treatment works (DWAF 2009b).

More generally, there is serious concern nationally about the competencies of personnel running municipal water services in many municipalities (Muller 2009). To a large degree, water services rest on public health engineering skills. But many municipalities have under-estimated the organisational requirements for effective water services provision, and municipal skills shortages are getting worse. In 2005, 83 municipalities employed no engineers at all, in any services (Lawless 2008). Four years on, the situation is far worse, with nearly 100 municipalities now employing no engineering professionals at all; overall, there has been a 10% fall in the number of municipal engineers since 2005, despite a significant expansion in service coverage. Just 44 municipalities out of 284 – 15% – employed registered professional engineers (Lawless 2009). Inappropriate appointments impact across the delivery chain, from planning, to design, to operational management, maintenance planning and so on, and impact on the ability of the technical services department to lobby successfully for adequate funding; collectively, these factors shape the quality of local service provision profoundly.

3.9.3 Limited Municipal Capacity to Deliver Water Services Effectively

Perhaps the most telling indicator is the finding of a October 2009 risk assessment compiled for DWA, which found that wastewater services in 23 municipalities were in a state of crisis, with an acute risk of disease outbreaks; a further 99 municipalities

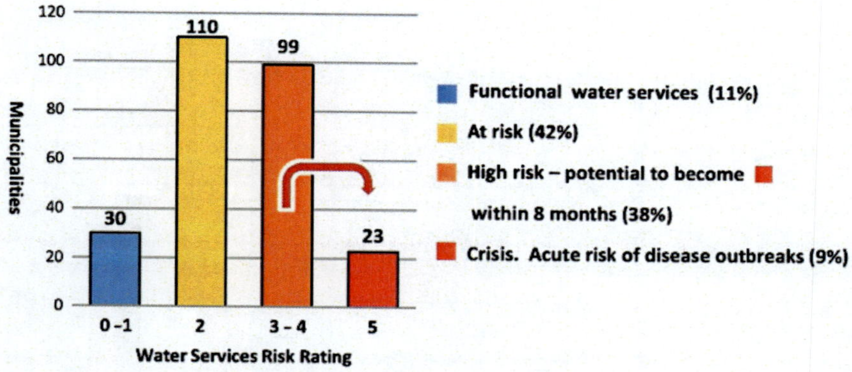

Fig. 3.2 Categorisation of municipalities by wastewater management risk rating

were at high risk of entering a state of crisis by mid-2010 if no turnaround is achieved (Fig. 3.2).

A further assessment in March 2010 confirmed the projected deterioration, and showed that the rate of decline was even greater than anticipated. The assessment looked primarily at technical and operational risk factors, and gave particular emphasis to compliance with wastewater treatment requirements. The list of municipalities at risk or high risk included the full spectrum of municipal categories, including two of the biggest metropolitan municipalities. The assessment is highly significant, because it counters the usual assumption that it is primarily the weakest municipalities, serving predominantly rural populations that face the gravest sustainability problems. Some of the most severe performance deficiencies are in secondary cities, with the most widespread problems found in municipalities serving predominantly small towns (CGTA 2009e; DWAF 2009b) (Fig. 3.3).

Whereas the greatest challenges in largely rural municipalities lie in providing access to basic services, largely through communal standpipes and VIP toilets, the challenge in more urbanised settlements relates to the more demanding technical challenges inherent in operating and maintaining reticulated water and sanitation services; at issue is the level of service provided. Compounding the O&M challenges are the higher operating costs and the need for more elaborate billing and collection systems, and the financial implications of providing high level services in a context of significant poverty and high unemployment. Under-investment in maintenance, refurbishment and expansion is eroding the ability of these municipalities to provide adequate services.

There is widespread evidence that MIG has funded the development of new infrastructure which is now incurring maintenance costs beyond what municipalities have budgeted for; widening cost-recovery gaps are undermining the ability of municipalities to fund infrastructure expansion, rehabilitation or renewals outside of low income areas, and compromising their ability to raise loans. Thus an unintended consequence of MIG in some municipalities is that it is starting to

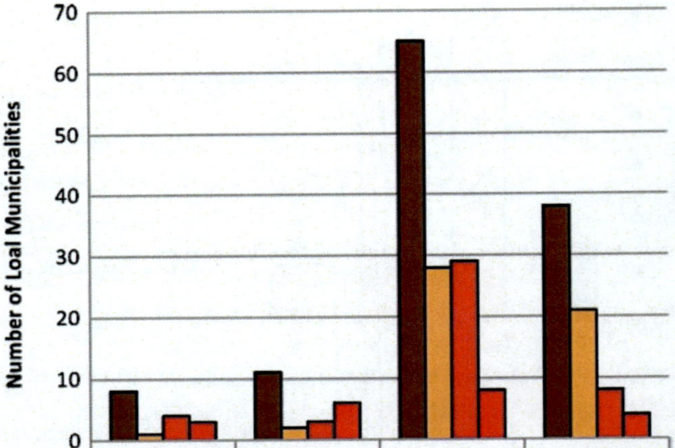

	B1	B2	B3	B4
Total scoring 3 or more	8	11	65	38
Scored 3	1	2	28	21
Scored 4	4	3	29	8
Scored 5 or higher	3	6	8	4

Fig. 3.3 Wastewater management risk rating by type of local municipality

impact on the state of overall economic infrastructure of municipalities; if services in the commercial and middle class areas deteriorate, they could cripple local economic development and the revenue base on which the municipality depends for wider service provision.

3.9.4 Growing Municipal Dysfunction

The Constitution assigns responsibility for delivery of basic services to local government, yet there is widespread evidence of severe dysfunction in the majority of municipalities. This impacts on municipalities' ability to plan and deliver services, and to manage their finances to support sustainable provision of services.

Structural weaknesses account for many of the challenges municipalities face: local government has been given an exceptionally wide-ranging and challenging mandate, yet many of the administrations are less than 10 years old, and have had to establish themselves as administrations in areas where it is hard to attract capable professionals; they face enormous challenges arising from decades of neglected investment and development, and an inadequate resource base to support the functions they must perform.

However, many of the challenges described above result from sheer mismanagement. This in turn is related to inappropriate appointments to meet employment equity targets, deployment of ANC cadres in senior positions for which they are not qualified and decision-making and resource allocation on the basis of party and factional politics rather than objective assessment of the requirements for sound performance (CGTA 2009e). These points have been made by the President, by Cabinet Ministers, senior government officials, and in a number of reports released in late 2009 which focused on the dire state of local government.

There is extensive evidence of weak financial management and poor management capacity in the majority of municipalities nationally. Analysis of municipal operating budgets for 2008/2009 shows that 24 municipalities overspent their operating budget to the value of US$350 million; more significantly, 166 municipalities *underspent* available funds by US$1.7 billion (National Treasury 2009b). This suggests that poor financial management is a greater challenge than a shortage of funds to provide services to people living in poverty.

There is a significant shortage of financial professionals in municipalities, and a high turnover of staff in all professional categories. Just one in five of 253 municipalities audited for the 2006/2007 financial year received an unqualified audit (Shiceka 2009). Without adequate financial planning and financial controls, cash flows are vulnerable, and a growing number of municipalities face bankruptcy and an inability to pay creditors. "A very significant risk going forward is that municipalities' spending plans outstrip realistically collective revenues," noted a major assessment of the state of local government in October 2009 (CGTA 2009d).

Worse, there is evidence of widespread corruption in municipalities, and abuse of political office and job status for personal gain (Ibid.) In August 2009, Yunus Carrim, Deputy Minister of the department responsible for local government, phrased the problem succinctly in an interview with a journalist (Carrim 2009a):

> We didn't anticipate the extent to which power struggles within municipalities would paralyse service delivery, and power struggles between local and district municipalities would undermine the two-tier model. And we didn't, just didn't, foresee the extent to which municipalities would become the soft underbelly of patronage and corruption in our country (Brown 2009).

The report of a national assessment of the state of local government noted in October 2009 noted that 'a culture of patronage and nepotism is now so widespread in municipalities that the formal municipal accountability system is ineffective and inaccessible to most citizens...There is now a lack of citizen confidence and trust in the system' (CGTA 2009d).

These realities underpin the surge of street-level protests across South Africa in 2009.

3.9.5 Protest Action

There were almost as many community protests in 2009 (105) as in the previous 5 years put together (Municipal IQ 2010); the number of protests doubled compared to 2008 (CGTA 2009d). A large part of the reason lies in the explanation given to delegates at a meeting of local government officials in mid-2009 by an advisor to government:

> Our communities have been neglected and regarded as passive participants in issues of governance. As a result, no dedicated effort has been given to societal structures in order to deepen democracy. Our people on the ground are taken for granted by service delivery points, thus leading to fierce confrontation between local authorities and communities; e.g. widespread riots in townships and villages (Nkontwana 2009).

In setting bold service delivery targets to overcome decades of under-development, government has created the expectation that local government can and will deliver. Faced with enormous challenges of scale and urgency, most municipalities have opted for a technocratic top-down approach to delivery, with decisions taken at high level and implemented with little public involvement or communication. But many municipalities lack the skills needed to succeed with a technocratic approach; in too many instances, the result has been non-delivery, or development of poor quality infrastructure without the resources to keep it working. Moreover, the approach has emphasized funding inputs and infrastructure outputs, not service delivery outcomes. Technology choices and implementation options have been limited, and options for customising service delivery options to address locally-identified problems and needs have often been overlooked.

Ward committees, established at sub-municipal level, are the primary forums through which public engagement is meant to occur. But ward committees are discredited and dysfunctional in large parts of the country. Councillors are frequently unresponsive – or unable to respond – to issues raised directly by citizens, fuelling growing frustration, mistrust and anger (CGTA 2009a).

Galvin and Habib (2003) note that South Africa has indeed decentralized delivery, but has opted for a highly state-centric approach, through municipal structures, rather than a more integrative and inclusive model which would enable it to harness the support and creativity of people in planning and sustaining services that meet their needs. To a large degree, ordinary people have become marginal to the process of municipal service planning and delivery, and, as the RDP warned presciently, have indeed become the objects of development (ANC 1994).

3.10 Towards a Turnaround

3.10.1 The Local Government Turnaround Strategy

Confronted by increasingly violent protest action, from mid-2009 the new administration under President Zuma undertook a wide-ranging assessment of the state of local government. Provision of decent services is the most important function of local government, yet the performance of many municipalities has been very poor. Facing national local government elections scheduled for early 2011, government has given priority to driving a comprehensive turnaround strategy for local government, with a strong emphasis on strengthening governance and accountability.

The approach was endorsed by Cabinet in December 2009, and within 3 months, by March 2010, every municipality in the country was required to draft its own turnaround strategy to remedy problems within its area of jurisdiction. The strategy has five core objectives:

- Ensure that municipalities meet the basic needs of communities, with systems to accelerate quality service delivery within the context of each municipalities conditions and needs
- Build clean, responsive, accountable local government
- Improve functionality, performance and professionalism in municipalities
- Improve national and provincial policy, support and oversight to local government
- Strengthen partnerships between local government, communities and civil society (CGTA 2009f)

Effective water services depend very largely on capable local government authorities, and thus the local government turnaround strategy has critical implications for a turnaround in water services. But the water sector faces very particular challenges of its own. What has been learned over the past 15 in South Africa?

3.10.2 Some Lessons from South Africa's Experience

If water services are a municipal responsibility, capable municipalities are a prerequisite for good services.

Across the infinite diversity of South Africa, municipalities which are doing well in delivering water service have good leadership, the right skills, effective communication strategies, sound financial management and strong accountability systems. Where the importance of any one of these is underestimated or neglected, the impacts are evident in service deficiencies.

One size does not fit all. Differentiated approaches are needed which recognise strengths, weaknesses and varied abilities across the entire spectrum of institutional structures. Institutional models for water services work best when they are developed on the basis of robust, comprehensive local assessment of what the key challenges are and how best to meet them.

Institutional restructuring is complex, risky, expensive and enormously disruptive to service delivery. Some form of institutional reform may be necessary to make better use of available resources in particular areas, but restructuring should not be the default response. Wherever feasible, the emphasis should be on rebuilding institutions, not restructuring. Many of the current weaknesses stem from poor accountability systems and inappropriate appointments, and can be remedied. There is considerable scope to make better use of the skills and competencies that are available, and to appoint people with the professional and technical competencies required to achieve effective and sustainable services.

Institutional continuity is vital for effective service provision, as it can take many years to develop the systems and competencies needed to deliver effective services. At an individual level, high turnover of staff is profoundly disruptive, because it takes several years for a new senior manager to build the knowledge base and experience needed to perform effectively on the job. Moreover, DWAF has invested enormous resources in support programmes to build the competencies of water services authorities, but the benefits are short-lived where key officials change jobs frequently.

Political representatives and service providers have distinct and separate roles. Councillors set the strategic direction and determine policies, and service providers are tasked with delivery of services in line with that strategic mandate.

The separation of Authority and Provider functions which underpins the draft regulatory strategy for water services has been achieved in only a minority of municipalities. However, DWAF's current emphasis on tracking key service outcomes – notably the incentive-based Blue Drop and Green Drop certification process, whereby municipalities that score highly across a range of requirements for good water and wastewater treatment – is delivering very positive results. This suggests that regulation is possible without a formal separation of WSA and WSP roles; what matters most are clear performance targets, sound operational procedures, effective performance monitoring and consequences for poor performance.

Outcomes are even more important than outputs. Enormous investment in water services infrastructure does not necessarily deliver sustained improvements in service delivery. There is substantial evidence of new infrastructure becoming dysfunctional within a very short time because of a lack of personnel with the requisite skills and competencies to operate and maintain that infrastructure. Equally, services with high running costs are being provided in many areas without adequate regard for the ongoing income required for viability. The combination of poverty, poor revenue management and inadequate funding support from the fiscus contributes significantly to service breakdowns.

Access to sufficient bulk water supplies, expansion and upgrading of bulk infrastructure and renewal of ageing networks is as necessary for improved service delivery as new taps and toilets. Many rural settlements continue to draw their water, untreated, from streams, because there has been no source development and regional bulk networks cannot meet demand; some housing projects have stalled, or have been abandoned and vandalised, because there is insufficient water to supply them, while the consequences of providing waterborne sanitation extensively without adequate water supply or waste treatment capacity is evident in growing supply interruptions and eutrophication of rivers. Upgraded services attract new settlement, with the potential to result in supply bottlenecks. In even the biggest metros with the best performing utilities, leaks and bursts from decayed pipe networks lead to pressure drops, water outages and compromised water quality. The scale of investment required vastly exceeds available funding.

The real costs of sustainable water provision are not reflected in the current pricing system, across the entire value chain from source to tap and back. There is no coherent pricing strategy for the water sector as a whole; DWAF's pricing strategy focuses on raw water, and each water board and municipality has its own approach (see Chapter 9). The result is a wide range of pricing approaches and tariff levels, with under-recovery of real costs and across the sector as a whole, and severe under-investment in asset management and renewal. One immediate indicator is the extent of leaks and losses resulting from decaying infrastructure, in a context of growing water demand and looming water scarcity. Two of South Africa's biggest cities, with the best-run water utilities, have Non-Revenue Water percentages well into the 1930s. This raises questions about the extent of real losses in municipalities with less sophisticated monitoring systems, and the quantum of funds required nationally to renew ageing networks – and instal comprehensive metreing systems.

Growing water scarcity requires a far stronger emphasis on water conservation and demand management to improve use efficiencies and cut avoidable losses.

Most municipalities do not ring-fence their income and expenditure for water services, and so do not know what it costs them to provide water services. Tariffs are frequently set at levels far below the real costs, and budgets are balanced largely by neglecting essential maintenance and forward investment. Compounding this, most municipalities use an accounting approach to tariff setting, based on balancing their annual budgets, rather than a financial approach which takes account of the real costs of long-term sustainable supply – which includes comprehensive asset management, with provision for rehabilitation, renewals and upgrading of infrastructure.

Growing grant dependence is masking significant inefficiencies in many municipalities. There is no question that many municipalities require additional funding support from the fiscus to provide decent services in a context of chronic poverty; however, where citizens and municipalities pay only a fraction of the real cost of services, their incentives for using water efficiently, reducing Non-Revenue Water and delivering services cost effectively are reduced. Water services need to be run more like professional business organisations – not to make a profit out of poor people, but to ensure that expenditure is matched by income, that cash flows are sufficient to fund operations throughout the year, that procurement needs are dealt with promptly, that provision is made to rehabilitate or replace depreciating assets and that there is a strong emphasis on providing services that meet people's needs.

Reliance on grants and transfers has shifted the focus of reporting and accountability away from the relationship between municipalities and the citizens they serve, to the relationship between municipalities and other spheres of government. It is imperative that a new relationship of mutual accountability is negotiated between councillors, officials and citizens, municipality by municipality, based on frank dialogue about where current challenges lie and what is needed to remedy them. Equally, the structure of government is premised on spheres, not tiers, of government; effective inter-governmental relationships require mutual accountabilities.

Service upgrading in informal settlements is linked to comprehensive settlement upgrading, but settlement development and housing delivery is progressing far more slowly than anticipated. One consequence is that millions of people live in limbo in informal settlements, with very uncertain futures: when will they get the free new house they have been promised? National housing policy advocates in-situ upgrading wherever possible, but in-situ development is often more complex and protracted than greenfields development, and thus does not provide fast-tracked solutions. Meanwhile, services in most informal settlements are extremely poor – and

38% of recent service protests stem from informal settlements (CGTA 2009d). There is an urgent need for incremental improvements in service provision in informal settlements, until such time that better housing can be developed.

Quick fixes seldom deliver lasting benefits. A range of interventions are needed to address challenges, and some can be implemented very effectively in the short term. But a long term perspective is needed to guide the range of complementary initiatives the sector needs to build a solid, sustainable base for effective service delivery.

South Africa's water services sector could be approaching a perfect storm, with consequences far more serious than those that have rocked the country in the electricity sector.

3.11 Strengthening Water Services

The need for a turnaround in municipal administration and governance is widely acknowledged. While this is necessary for an improvement in water servicing, it is not necessarily sufficient. Equally, there are questions about how best to achieve a turnaround in water services, in a context where water services delivery is inextricably linked to municipal powers and functions. Regardless of how well a water services provider performs operationally, in the South African context it relies on the municipality for billing and revenue functions, and it is the municipality that sets the tenor of the utility's relationship with its customers.

On the other hand, turning around a municipality takes time, and immediate interventions are needed in many municipalities to strengthen water services provision. Take the case of Emfuleni Local Municipality, on the banks of the Vaal River south of Johannesburg. For over a decade there have been severe and worsening sewage spills into the river, with total faecal coliforms – a measure of sewage contamination – exceeding 4,000,000/100 ml at one stage (Rand Water 2002–2009). The reasons stem largely from municipal mismanagement. One remedy proposed in the early 2000s was to separate out the water services function from the rest of the municipality, and establish a dedicated municipal-owned entity, like Johannesburg Water (Pty) Ltd, to run water services. National Treasury opposed this, on the grounds that it would be better to turn around the entire municipality. Enormous resources have been invested in municipal support and restructuring initiatives, but the results have been disappointing, with little significant improvement in the performance of the municipality or its wastewater management.

The following sections reviews some options for strengthening water services provision.

3.11.1 Technical Support

There have been a number of valuable technical assistance programmes in recent years which target municipalities and their service functions. One prominent initiative is Siyenza Manje, run by the Development Bank of Southern Africa, where professionals are assigned to struggling municipalities to provide short-term support. To date, over 600 professionals have been deployed in municipalities around the country. However, this relief is insufficient to achieve lasting change, and some who have worked on the Siyenza Manje programme complain that municipalities do not necessarily wish to follow their advice or act on the evidence of corruption they find, and that local politicians frequently overrule the decisions of officials; moreover, some municipalities claim that they have lost technical staff to the Siyenza Manje programme, causing problems where previously they did not exist.

Technical assistance is all about inputs; what is needed is greater debate about how best to achieve the outputs and outcomes that are needed, with a far stronger emphasis on improved performance management.

3.11.2 Professionalisation of Water Services Provision

Achieving a lasting improvement in performance outcomes requires institutional reforms, and greater professionalisation of municipal water services. This ranges from recruiting the necessary skills and competencies, to a greater emphasis on performance contracts and service delivery outcomes, to ring-fencing the finances and ideally administration of the water services function, through to the establishment of separate utilities in some cases.

An obvious requirement is to insist that municipalities appoint personnel with the appropriate professional and technical expertise, whether at management or supervisory level, or through appointing works operators with the skills specified in sector regulations. DWAF is considering introducing mandatory minimum qualifications for senior water services personnel, possibly with accreditation through a professional body (H. Muller, 2009a, personal communication, 10 December 2009; M. Muller 2009).

3.11.3 Regulation

Strong sanctions for weak performance are a powerful driver of institutional reforms, but water services regulation is still evolving. The Department of Water Affairs is the sector regulator, but its emphasis until recently on "developmental regulation" – i.e. supportive engagement with municipalities to remedy problems – blurred the line between assistance and enforcement (or even intervention). Further, the Constitutional principle of cooperative governance between the three spheres of government has left the national sphere reluctant to intervene decisively at the local sphere, let alone prosecute municipalities who fail in their service obligations. One consequence is that there are few sanctions for municipalities who do not comply with sector legislation. There is evidence of DWAF taking a tougher stance, through issuing directives to a growing number of municipalities and requiring them to formulate action plans to remedy non-compliance; court action is currently a last resort, and seldom used. In practise, however, court action solves very little unless individuals are held accountable for poor performance, and to date no individuals have been prosecuted.

Given the extent of the crisis in municipal water services, it is very likely that DWAF's regulatory powers will be strengthened, with a greater emphasis on enforcement and penalties for municipalities that do not comply. Further, acknowledgement of the crisis has opened a space to review the powers and functions of a municipality where it is not able to perform its function, with the possibility of mandatory restructuring its water services function; alternatively, it could be compelled to bring in a professional service provider. Without strong sanctions for poor performance, the incentives for a turnaround are limited.

There is, nonetheless, heartening evidence of better water quality management, spurred by regular monitoring and publication of municipal performance. An electronic Water Quality Management System tool, eWQMS, is now being used by every Water Services Authority in the country, with an average of 90% of municipalities submitting data each month. And in mid-2009, DWAF awarded the first Blue Drop and Green Drop certificates, to municipalities achieving at least 95% and 90% compliance, respectively, across a number of weighted performance criteria for management of water and wastewater quality. DWAF believes that introducing simple indicators for two key performance areas will ease the regulatory burden on the Department (Manus 2008).

Until recently, every municipality in the country liked to claim that the water they supplied complied with the requisite treatment standards, but far fewer could provide the data to prove it; in 2005, just 42% of Water Services Authorities were monitoring drinking water quality, but this had grown to over 90% by 2008 (DWAF 2008), spurred by the introduction of eWQMS (electronic Water Quality Management System), an open-source, web-based tool for capturing, disseminating and managing data on water and wastewater quality. The tool, introduced nationally from April 2006 after application initially in the Free State province, came with extensive DWAF-funded hands-on support to municipalities to help them to collect and capture the data and respond to the results. The tool enables municipalities to load the results of all tests onto the web each month, and draw reports on the results which flag compliance with mandatory standards. The data is forwarded directly to DWAF, as sector regulator (Mackintosh et al. 2008).

The results have been dramatic. Weaknesses have been revealed in many areas, but far more important is the impetus and encouragement given to municipalities to measure and track their performance on a monthly basis. The layout of eQWMS reports guides municipal technical staff in ensuring that all basic water quality management requirements are prioritised and implemented.

eWQMS currently focuses on large-scale municipal water distribution systems, but the approach is now being extended to address the thousands of small localised boreholes which fall outside of mainstream water quality management systems. With support from the World Health Organisation and the Gates Foundation, work is now being done on the development of simple test kits that can be used in remote areas to assess free chlorine levels, turbidity, pH and the presence of faecal coliforms.

The aim is to use mobile phone technologies to input the data directly to a central database from where it can be assessed and acted on (Moorgas et al 2009).

Of course, eWQMS cannot compensate for inappropriate staffing or lack of funds for system rehabilitation; regulation is a more appropriate lever for institutional reform or budget re-prioritisation. However, eWQMS has provided the information management and reporting system needed at municipal level to enable DWAF to regulate water quality nationally. It has also laid the basis for the introduction of the Blue Drop and Green Drop regulatory system from 2008, which aims to affirm good performance, and mobilise public pressure in municipalities which do not meet compliance requirements. The first round emphasised a 'name and praise' approach, with public affirmation of the 22 authorities which met the 'Blue Drop' standard; poor performance at the vast majority of wastewater treatment works overshadowed the achievements of the 32 works (out of more than 800) that were awarded Green Drop certificates in 2009.

3.11.4 Alternative Institutional Models

South Africa's experience in setting up external, stand-alone water utilities has been mixed. The experience of Johannesburg Water (Pty) Ltd, wholly owned by the City of Johannesburg Metro Municipality, and supported for 5 years through a very productive management contract, illustrates the benefits of setting up a professional, commercially-oriented utility. These include the ability to focus on water services, provide a potentially more attractive work environment, greater flexibility in recruitment and procurement, and a clearer separation between political and technical functions. However, there are also significant risks and costs as experiences elsewhere have shown. Moreover, the experience of Ilembe District Municipality, which uses a private sector concessionaire in one area, shows the critical importance of strong capacity in the municipality to manage and monitor its service contract with the utility and support effective communication with customers (Galvin 2009); even with outsourcing, responsibility for ensuring good services remains with the municipality.

To date there has been very little discussion of the role of non-municipal role-players other than water boards following a backlash against a pro-privatisation lobby in the 1990s. Privatisation, through the sale of public assets required for the delivery of basic services, is explicitly prohibited in the 2003 Municipal Finance Management Act, and most municipalities and the labour movement are strongly opposed to delegating management of water services to non-public service providers.

A growing number of municipalities are drawing on Water Boards, or regional bulk water providers, to supplement their services capabilities. There is an important role for water boards in supporting service delivery by municipalities, where they have the necessary retail servicing competencies to do so; however, the capabilities of water boards vary widely and some are facing significant performance and sustainability challenges of their own. Water Boards are a particularly attractive support option and partner for many municipalities, because competitive tender procedures are not required when appointing public entities; this makes it easy for municipalities to bring in supplementary capacity fairly quickly. However there are growing questions about whether this is the most effective approach, and whether municipalities (and tax-payers) are getting adequate value for money.

In municipalities serving people living primarily in rural settlements, there is some acknowledgement of the need for local delegated management arrangements which involve local civil society groupings working with small and medium enterprises (CGTA 2009b). Welcome and necessary though this approach is, it does not address the need for a step change in South Africa's approach to managing water services. Providing good quality water services is far more complex than is generally acknowledged. It requires considerable technical and managerial expertise, and a firmer emphasis on achieving service outcomes that are sustainable and affordable to the country as a whole.

Increasingly international debates are moving beyond a polarisation between public-sector and private-sector led approaches, to discussion of what the nature of partnering relationships should be, and how to make them work to serve those most in need of services. This is not to suggest that the answer to South Africa's water services problems lies in contracting the private sector; the current challenges are far more complex and diverse than a simple resort to market-driven approaches could remedy. Moreover, there are examples of good and bad private sector entities, just as there are examples of good and bad public sector entities. At issue is how available resources – human and financial – are mobilised and configured, and what kinds of partnerships might prove effective in different contexts – perhaps public-public or public-private outcomes-based management contract, or a public-private partnership with delegated management of a publicly-owned treatment works, or a public–public-CSO or public–private-CSO model to strengthen service delivery in rural settlements located some distance from the municipal administrative hub, as proposed in a recent MIG review (CGTA 2009b) and so on; and how service providers are held to account to ensure that the needs of all users – particularly those of the poorest and most vulnerable citizens – are met.

3.11.5 Improved Financial Sustainability

Government faces infinite demand from municipalities for funding support for capital and operating expenditure, because of a combination of mismanagement, poor revenue collection and the need to provide subsidies for households receiving services at a level they could not otherwise afford. Greater professionalisation of municipal water services provision, with improved performance efficiencies, is essential, because the cost of the current approach is increasing unsustainable. Growing climate variability will add a range of further management costs, and the combination of growing water scarcity and deteriorating water quality means that containing costs and keeping water affordable to all users is imperative.

Municipalities are looking to government to fund water services infrastructure development to a degree not envisaged in any existing funding frameworks. MIG funded an estimated US$2.9 billion in water services infrastructure between 2003–2009 (DWAF 2009a), but Eastern Cape municipalities told the Minister of Water Affairs in July 2009 that they need US$5.6 billion in their province alone to repair and rehabilitate ageing water and sanitation infrastructure, with similar amounts requested from Limpopo (Hennop 2009). Without improved investment in maintaining existing infrastructure, the amount required for rehabilitation will grow. DWAF, meanwhile, estimates that US$10.8 billion is required to address bulk water infrastructure needs in municipalities (Van Zyl 2009), with at least US$3 billion needed for refurbishing and upgrading wastewater treatment works.

Funding from the private sector is an important source of additional capital funding for municipalities, but banks are wary of lending money where the risks are high, where they are not confident of a municipality's ability to repay a loan and where they have little confidence in current service provider mechanisms. Sourcing the additional funds that municipalities need – particularly for essential infrastructure development that falls outside the scope of MIG – needs to start with effective institution building. Improved billing and collection by municipalities is imperative, but with water tariffs set at sub-economic rates in most municipalities, even improved collections will be insufficient. Tariff increases raise concerns about the affordability of water to consumers already reeling from electricity price hikes. More immediately, water users will resist price increases unless they come with evidence of significant improvements in service quality.

As access to services has expanded, so too has the cost to municipalities of operating and maintaining those services. Municipalities planned to spend US$1.76 billion on water and sanitation in 2009/2010, compared to US$0.92 billion in 2003/2004; operating expenditure has doubled in just 6 years (National Treasury 2009b). Concurrently, municipal dependence on grants to fund their operations is growing. The share of service charges in the total operating revenue of local government declined from 49%t in 2003/2004 to 42.9% in 2009/2010, mainly due to the sharp increase in national transfers (Carrim 2009b), but also as a result of what National Treasury describes as 'a lack of fiscal effort' – i.e. poor billing and collection – on the part of municipalities (National Treasury 2008). In 2008, 57 municipalities received more than three quarters of their revenue from national transfers, and at least six relied on grants for 100% of their income (CGTA 2009e). Municipal debt soared to US$7.16 billion by mid-2008, with 60% of that amount owed by households (EU Municipal Outreach 2009).

A further area of financial vulnerability for municipalities relates to the provision of Free Basic Water, which provides important support to poor and vulnerable households in line with government's pro-poor policy objectives. The subsidy amounts to between US$3.5 and US$8 per household per month, depending on the volume of free water provided and the local real cost of provision. DWAF statistics suggest that 85.6% of households country-wide enjoyed access to free basic water in 2009 (DWAF 2009a); demand-side figures collected by Statistics South Africa suggest far lower coverage (Stats SA 2009a).

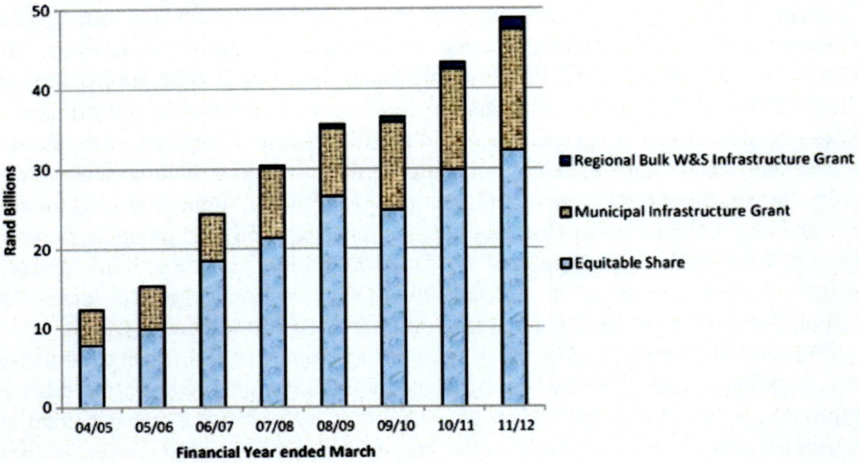

Fig. 3.4 Grants and transfers to municipalities, 2004/05–2011/12

The original intention of the free basic water policy was to provide relief to poor households, but in practise it is extremely difficult to target poor households alone. But blanket application of Free Basic Water policies is becoming increasingly problematic in many municipalities where the cost of providing free basic water is greater than the resources allocated to fund it. In many areas underfunding is contributing to the deterioration of services; deteriorating services are decidedly anti-poor.

Many factors contribute to this. Through the Equitable Share, National Treasury allocates substantial funding intended to cover the cost of a quantum of basic services to poor households, and transfers this directly to municipalities; municipalities do not necessarily use the Equitable Share to fund basic services to the extent required, and commonly absorb a large portion of the transfer into general administrative expenses, leaving a shortfall on the cost of providing free basic services.

A more complex problem is that Treasury assumes that municipalities can target the allocation of Free Basic Services to defined indigent households, and provides funding support accordingly. Treasury uses extrapolated 2001 census data to determine the number of indigent households in each municipality, defined at a relatively low income threshold, and allocates Equitable Share funds per municipality on that basis (National Treasury 2009a). But municipalities have their own definitions of indigency; Cape Town, for example, sets its threshold for qualifying households at a monthly income level that is three times higher than Treasury's. This immediately results in a short fall between the funds available through the Equitable Share, and the number of households the municipality regards as eligible.

Moreover, municipalities cannot target poor households as readily as Treasury can quantify them. Targetting tends to rely on means-testing and municipal indigency registers, but these are onerous and very problemmatic, with substantial errors of exclusion and inclusion. Consequently most municipalities opt for administrative pragmatism and universalism; water from communal standpipes is provided free and unmetered, and elsewhere, municipalities use a rising block tariff, with the first tariff block set at zero for everyone. In theory, the cost of providing a portion of water free should be offset by cross-subsidies from tariffs charged in higher consumption bands. In practise, many municipalities do not charge a volumetric tariff, and thus the monthly service fee is unrelated to actual consumption. Even where municipalities do charge a volumetric tariff, very few beyond the six metropolitan municipalities have sufficient large volume paying customers to offset the cost of universal provision of free basic services; even fewer are implementing volumetric tariffs at the level required. Consequently there is a significant disjuncture between what the Free Basic Water policy aims to achieve, how the Equitable Share aims to support it, and what municipalities can implement practically on the ground. In short, sustainable provision of free basic water requires either significant local cross-subsidies, or a degree of administrative sophistication that is simply beyond the reach of many municipalities.

Refocusing free basic services to target a far narrower band of households will help to close the gap between income and expenditure, using currently available funding; but this is likely to prove extremely costly in political terms, and targetting raises complex challenges around exclusions. Other alternatives could tap funding sources beyond water tariffs, such as property rates, to support more viable universal provision, or link eligibility and benefits to social grants. There are no easy answers; but what is clear is that a different approach is increasingly necessary to safeguard the policy's pro-poor intent.

3.11.5.1 The Phiri Court Case

In Johannesburg, the metro targeted Soweto for the first large-scale introduction of prepayment metres from 2004, because non-revenue water levels there were 67%, compared to about 20% in metered areas (Marin et al. 2009). Payment levels were extremely low – around 10% – and the extent of physical and commercial losses was unsustainable. Installation of the metres occurred in the context of an extensive infrastructure upgrading programme to augment supply and improve water pressure. Opponents of prepayment metres took the City to court, arguing that prepayment metres were unconstitutional, and that 'silent disconnections' when the free basic allocation, or available credit was exhausted, compromised people's right of access to water. The Constitutional Court ultimately affirmed the Constitutionality of prepayment metres, provided that a reasonable amount of free basic water is provided. However the court case has had the effect of discrediting prepayment metres and legitimating non-payment for services, and metre bypasses and illegal connections have soared in Soweto. In June 2009 the city wrote off R2.8bn in bad debt, in what was the third major debt write-off in 8 years; steadily rising non-revenue water and water services bad debt contributed significantly to this amount.

Free Basic Water is a bold and important support measure, but it has no exit strategy and faces only growing demand; and with one in four South Africans now receiving a social grant, the cost of current poverty relief measures is vast and growing. A far-reaching shift in public perceptions of the value and cost of water is necessary if the sector is to avoid financial collapse. Illegal connections and metre by-passes are widespread, and service payments levels are far below what is needed to support good service provision. A more effective approach to communicating the challenges facing the sector is needed; building a culture of co-responsibility for service provision, premised on payment for services beyond the free basic amount, will require little short of a new social compact between government and citizens.

3.11.6 Building an Active Citizenry

Current debate about the underlying reasons for inadequate service delivery creates an important opportunity for South Africa's people to reflect on how they have framed their relationship with government, and to explore new options for re-creating the relationship between government and citizens, both to strengthen democracy and to strengthen service provision. If citizens' role is only to demand services and accountability from government, then they have rights with no reciprocal responsibilities (Mathekga and Buccus 2006; Nemeroff 2006). If the roles of citizens and government are more balanced, officials and elected representatives may be held more accountable, and citizens can make a greater contribution to good servicing; examples of their role ranges from reporting bursts and leaks, to paying for services, to not condoning water theft, illegal connections and metre bypasses, to shaping local planning and service improvements. A new partnership or social compact is needed between government and citizens. An important starting point is local dialogue, where politicians, officials and citizens from different constituencies come together to build understanding of the reasons for evident problems and jointly formulate strategies to address them (Galvin 2009).

Good governance involves more than government; this fact is acknowledged in the draft Water Services Regulation Strategy, which makes reference to the need for additional local 'Citizens' Voice' initiatives which provide users with a forum to engage the municipality and its service provider around service concerns (DWAF 2007a). These user platforms have the potential to strengthen both local accountability mechanisms and the effectiveness of regulation at a national level, while also giving the municipality valuable feedback on areas needing urgent attention. A prerequisite is a structured local education programme, which takes both citizens and councillors through a series of topics – the water cycle, the national and local policy environment, the water services rights and responsibilities of citizens, and so on; this input provides a foundation for understanding where and why problems arise, and how to remedy them. Citizens' Voice initiatives have been introduced in Cape Town, Ekhurleni, eThekwini and uMsinduzi, with strong demand from a growing number of other municipalities (WIN-SA 2009).

Water sector-specific initiatives like Citizens' Voice are vital, and must be complemented with broader recourse measures – not least to assist citizens burdened with incorrect billing. But the far larger challenge is to reconstruct an engaged and active citizenry, where all the country's people feel their voices are heard and respected. Given the extent of joblessness, experienced most acutely by the youth, and the structural nature of absolute and relative poverty in South Africa, this is no easy task.

3.12 Conclusions

Fifteen years on from 1994, South Africa is a vastly different country. The challenges facing South Africa today are far more complex than in the heady early days of liberation – competitiveness in the global economy, chronic unemployment, AIDS, structural poverty, skills gaps, electricity shortages, corruption, a breakdown in trust in local government, and so on. For far too many people, tangible change has not yet come to the degree they had hoped for, and their patience is wearing thin. Improved service delivery is non-negotiable, yet achieving it has proved to be far more complex than the country's new leadership imagined.

Reforms in the water services sector have been implemented in parallel with far-reaching reform and restructuring of local government. Target-driven service delivery has focused on infrastructure, not people, and on capital projects, not sound operation and maintenance. The emphasis on meeting a critical national imperative – universal access to water services – has, to some extent, undermined the logic of decentralisation. Responsibility for delivery has been devolved to local government, but municipal accountability for delivery is to the centre, and to provincial and national leadership. The pressure on municipalities to meet targets and absorb large annual grants has substantially narrowed the scope for meaningful local civic engagement in deciding local needs, priorities and options, and building a sense of shared responsibility for the outcomes.

Good management of municipal water services is vital for social justice, growth and development and national water security. Climate change and unpredictable rainfall introduces new risks and uncertainties and raises the stakes. Meeting these challenges calls for water services institutions that are competent to deliver reliable supplies of safe drinking water in a context of deteriorating raw water quality, rising demand and growing scarcity, and which can manage the collection and safe treatment of wastewater without imposing harsh burdens on future generations. These issues underline the importance of cost-effectiveness and affordability in water services, nationally, to safeguard the rights of access of the most needy and vulnerable citizens, and to contain the cost to the state in a context of high and rising subsidies.

This chapter has reviewed government's performance in acting positively to honour the Constitution's promise of a right of access to adequate water for all. Service provision has become strongly state-centric; appointments are often shaped more by affiliation than competence; and decisions are taken too often on the basis of factional politics rather than professional assessment. Government is providing water services to an increasingly alienated citizenry, pursuing technocratic approaches which require a strong skills base and growing subsidies. In many areas, the nature and quality of the services provided fall far short of what citizens believe is reasonable. Widespread evidence of failures in water services delivery creates an important opportunity for South Africa's people to reflect on how they have framed their relationship with government, and what power they have handed over. Some form of new partnership or social compact is needed between government and citizens, to strengthen both service delivery and democracy, and to harness the willingness and the commitment of the great majority of people, who want to contribute to making things work better.

References

Africa E (2009, June) Department of Cooperative Government and Traditional Affairs Strategic Plan & Budget. (presentation to the Parliamentary Portfolio Committee on Provincial and Local Government). http://www.pmg.org.za/report/20090630-department-co-operative-governance-and-traditional-affairs-strategic-

ANC (1994) The Reconstruction and development programme. African National Congress. http://www.anc.org.za/rdp/rdp.html. Accessed 10 Oct 2007

Armstrong P, Lekezwa B, Siebrits K (2008) Poverty in South Africa: A profile based on recent household surveys. Stellenbosch Economic Working Papers: 04/08. http://ideas.repec.org/p/sza/wpaper/wpapers52.html.

Bhorat H, van der Westhuizen C, Jacobs T (2009) Income and Non-Income Inequality in Post-Apartheid South Africa: What are the Drivers and Possible Policy Interventions? Development Policy Research Unit, University of Cape Town, Working Paper 09/138, August 2009.

Benson, F (2009, February 09) Overview of the Cholera Outbreak. Department of Health.' Briefing to the Parliamentary Portfolio Committee on Water Affairs. http://www.pmg.org.za/report/20090204-department-health-department-water-affairs-forestry-responses-cholera

Brown K (2009, August 22). 'SA's provinces under fire', The Weekender

Carrim Y (2009a) 'Addressing the financial challenges in municipalities in the context of a review of the local government model.' Address to the Institute of Municipal Finance Officers Annual Conference, Johannesburg, 6 October 2009. http://www.cogta.gov.za/index.php?option=com_content&task=view&id=587&Itemid=36

Carrim Y (2009b, November 09). Address to the DBSA Knowledge Week

Cartwright A (2003) Infrastructure and local government: what role for the new Municipal Infrastructure Grant in South Africa? Hologram Newsletter. http://www.ksp.org.za/holonl29.htm

Daily Dispatch (2009, July 21). http://blogs.dispatch.co.za/dispatchnow/2009/07/21/crisis-in-mthatha-your-

Department of Co-operative Governance and Traditional Affairs (2009a, July 24). Ministerial task team on service delivery protests in Mpumalanga presents preliminary report. http://www.thedplg.gov.za/index.php?option=com_content&task=view&id=565&Itemid=30

Department of Co-operative Governance and Traditional Affairs (CGTA) (2009b). MIG Policy Review. Draft Discussion Document

Department of Co-operative Governance and Traditional Affairs (CGTA) (2009c). Basic Services Publication. Comparative Information on Basic Services, 2009

Department of Co-operative Governance and Traditional Affairs (CGTA) (2009d). The State of Local Government Report

Department of Co-operative Governance and Traditional Affairs (CGTA) (2009e, November). Local Government Turnaround Strategy. Working Together, Turning the Tide in Local Government

Department of Co-operative Governance and Traditional Affairs (CGTA) (2009f, December 03). Cabinet approves Local Government Turnaround Strategy. http://www.cogta.gov.za/index.php?option=com_content&task=view&id=598&Itemid=30

DPLG (Department of Provincial & Local Government) (2001, November 22). Ministerial Advisory Committee on Local Government Transformation, 'Interim report on the challenges facing local government', (submitted to the Minister of Provincial and Local Government)

Department of Health (DOH) (2008) 'Report on Childhood Deaths, Uhkahlamba District, Eastern Cape. http://www.health.org.za/documents/7f3d48106e60e6e5d164698624c5efb3.pdf?PHPSESSID=1a9e0a0951. Accessed 30 April 2008

DWAF (2007a) National Water Services Regulation Strategy. Draft

DWAF (2007b) 'Delmas diarrhoeal outbreak briefing', Briefing to the Parliamentary Portfolio Committee on Water Affairs. http://www.pmg.org.za/minutes/20071120-delmas-diarrhea-outbreak-2007-department-briefing. Accessed 21 Nov 2007

DWAF (2008, 4 June). Drinking Water Quality in South Africa 2008. Presentation to the Portfolio Committee Public Hearings on Water Quality. www.pmg.org.za/files/docs/080604dwaf.ppt. Accessed 14 July 2009

Department of Water Affairs and Forestry (DWAF) (2009a) Annual Report. 1 April 2008 to 31 March 2009

Department of Water Affairs (DWAF) (2009b) *Under-capacitated municipalities as at 30 October 2009.* unpublished internal document.

EU Municipal Outreach (2009, July 03). 'Who owes municipalities?' Newsletter 28. http://www.eumunicipaloutreach.org.za/newsletters/newsletter-28-2013-who-owes-municipalities-2013-3-july-2009/

Galvin M (2009) Straight Talk to Strengthen Delivery in the Water Services Sector. The Water Dialogues-South Africa Synthesis Report

Galvin M, Habib A (December 2003) The politics of decentralisation and donor funding in South Africa's rural water sector. J S Afr Stud 29:4

Groenewald Y (2009) Eastern Cape 'on the brink of a water crisis'. Mail & Guardian, 12 June 2009

GTZ (1998) Mvula Trust – A case study of an independent approach to rural water supply and sanitation in South Africa. http://www2.gtz.de/ecosan/download/mvulatrust-casestudy-southafrica.pdf

Hennop J (2009) 'Officials whipped at Water Indaba', 30 July 2009 http://blogs.dispatch.co.za/dispatchnow/2009/07/30/officials-whipped-at-water-indaba-map/

Johannesburg W (2003) Business Plan 2003/04

Kok P, Gelderblom D, Oucho J, Van Zyl J (2005) Migration in South and Southern Africa. Dynamics and determinants. HSRC Press, Cape Town

Lawless A (2008) Numbers and Needs in Local Government

Lawless A (2009, October 28) Presentation to the Local Government Indaba. Birchwood Hotel, Boksburg

Mackintosh G, Wensley A, Streuders K, Stevens F, Moorgas S (2008) 'Use of a performance management tool for effective municipal water quality management in South Africa', paper prepared for the IMESA Conference, October 2008, Bloemfontein.

Marin P, Mas J-P, Palmer I (2009) Using a Private Operator to Establish a Corporatised Public Water Utility. The Management Contract for Johannesburg Water. Water Working Notes, No. 20, World Bank, June 2009

Mathekga R, Buccus I (2006) The challenge of local government structures in South Africa: securing community participation. IDASA. Centre for Public Participation. http://www.idasa.org.za/gbOutputFiles.asp?WriteContent=Y&RID=1717

Manus L (2008) 'Drinking Water Quality in South Africa'. Presentation to the Portfolio Committee Public Hearings on Water Quality. http://www.pmg.org.za/files/docs/080604dwaf.ppt. Accessed 6 June 2008

Medical Research Council (MRC) (2009) Time for task-shifting: 999 days to close the HIV/AIDS treatment gap. http://www.msf.org.za/viewnews.php?n=275. Accessed 3 April 2009

Moorgas S, Naidoo V, Charles K, Mackintosh G (2009) Enhancement of the existing eWQMS for a sustainable water services programme in KwaZulu Natal', paper prepared for the bi-ennial WISA Conference, Durban, April 2010

McLennan A, Munslow B (eds) (2009) The politics of service delivery. Wits University Press, Johannesburg

Muller M (2007) Parish pump politics: the politics of water supply in South Africa. Prog Dev Stud 7:1

Muller M (2009) Towards the regulation of the competencies of South Africa's Water Services Managers. June 2009. WRC Report No TT 401/09

Municipal IQ (2010) Service delivery protests: first quarter 2010. http://www.municipaliq.co.za/press/press_release_service_delivery_protests_update_q1_2010.doc

Municipal Outreach (2009) Why do most municipalities fail to deliver? Newsletter 29. http://www.eumunicipaloutreach.org.za/newsletters/newsletter-29-2013-why-do-most-municipalities-fail-to-deliver-2013-10-july-2009. Accessed 10 July 2009

National Treasury (2008) The 2008 Local Government Budgets and Expenditure Review 2003/04–2009/10.

National Treasury (2009a). Explanatory memorandum to the division of revenue. Website annexure to the 2009 Budget Review. http://www.treasury.gov.za/documents/national%20budget/2009/default.aspx

National Treasury (2009b, August 28). Press Release: 2008/09 Municipal Budget Information: Fourth Quarter Financial Results as at 30 June 2009. http://www.treasury.gov.za/comm_media/news_archive_2009.aspx

Nemeroff T (2006) Probing the Protests. DeliverySA, October 2006

Nkontwana N (2009, July 15) Presentation of the New Mandate of Cogta. Birchwood Hotel, Boksburg. http://www.thedplg.gov.za/index.php?option=com_docman&task=cat_view&gid=117&Itemid=27

Philip K, Hassan EK (2008) Review of Second Economy Programmes: 15 Year Review. Second Economy Strategy Project

Pithouse R (2009, July 23) Burning message to the state in the fire of poor's rebellion. Business Day

Presidency (2009) Development Indicators 2009. http://www.thepresidency.gov.za/main.asp?include.../indicators/2009/index

Rand Water (2009). Rand Water Quarterly Reports, 2002–2009. http://www.reservoir.co.za/catchments/vaal%20barrage/barrage%20reservoir%20forum/barrage%20reservoir%20reports.htm

Republic of South Africa (1994) White Paper on Water Supply and Sanitation

RSA (2003) Strategic Framework for Water Services

SAPA (2009) 'River water tested after eight die', 8 June 2009. http://www.int.iol.co.za/index.php?set_id=1&click_id=125&art_id=nw20090608161915143C631531. Accessed on 14 December 2009

SAPA (South African Press Association) (2010) 'Sewerage problems a challenge not a crisis – Sonjica'. Press report, 29 April 2010. http://www.politicsweb.co.za/politicsweb/view/politicsweb/en/page71627?oid=173566&sn=Detail

Republic of South Africa (1998) White Paper on Local Government

Savage D (2007) Key themes and trends in municipal finance in South Africa. In: Donk V et al (eds) Consolidating developmental local government. Lessons from the South African experience. UCT Press, Cape Town

Savage D, Timm J (2003) Review of Water Services Delivery Strategies in South Africa, 1994–2003 (Unpublished report)

Schmitz T (1999) Rethinking delivery? A review of the efforts of the Department of Water Affairs, 1994–9. Centre for Policy Studies. Policy Brief 16. http://www.cps.org.za/16.htm

Shiceka S (2009) Address by the Minister for Cooperative Governance and Traditional Affairs, Honourable Sicelo Shiceka, on the launch of the 2014 Clean Audit Project, Birchwood Hotel, OR Tambo Conference Center, Boksburg. http://www.info.gov.za/speeches/2009/09071716251002.htm. Accessed 14 July 2009

Sonjica B (2009) National Assembly: Question 186 for Written Reply. http://www.dwaf.gov.za/communications/Q&A/2009/Q186June09.pdf. Accessed 14 July 2009

Statistics South Africa (2008) Income and expenditure survey 2005/06

Statistics South Africa (2009a) Non-Financial Census of Municipalities for the year ended 30 June 2008

Statistics South Africa (2009b) 2008 General Household Survey

UNDP (2009) Human Development Report 2009. http://hdrstats.undp.org/en/indicators/91.html

UNICEF (2009) The State of the World's Children. www.unicef.org/sowc/

Van Zyl F (2009, August 07). Input to hearings on the Finance and Fiscal Commission recommendations on the Division of Revenue 2010/2011. http://www.pmg.org.za/report/20090807-public-hearings-financial-fiscal-commission-ffc-recommendations-divis

Water Information Network (WIN-SA) (2009) Public Accountability through 'Citizens' Voice' – City of Cape Town Shares Good Practice, Lessons Series 20

Yeld J (2009) 'E Cape rivers tainted with human faeces'. 31 July 2009. http://www.iol.co.za/index.php?set_id=1&click_id=143&art_id=vn20090731103923270C109613. Accessed on 14 December 2009

Chapter 4
Water, Sanitation and Wastewater Management: Some Questions for National Water Security in South Africa

Kathy Eales

Abstract In South Africa, flush toilets have historically been associated with white privilege, and dry toilets with racial discrimination. The sector slogan 'Sanitation is Dignity' has deep resonance, and in urban areas, anything other than a flush toilet is regarded as inferior and at best an interim option. The emphasis on dignity, rights and aspirations has led to a focus on toilet technologies, rather than on integrated sanitation improvement. But rapid extension of reticulated water and sanitation infrastructure is creating significant bulk infrastructure bottlenecks; the focus on meeting coverage and bucket eradication targets far bolder than the MDGs is compromising sound operation and maintenance, and there is widespread evidence of wastewater treatment failures which have severe consequences for human health and the natural resource base.

This paper maps some challenges around service provision and sanitation improvement in urban South Africa. It argues that a rights-based approach to providing water and sanitation is a hollow promise if the service is unsustainable and prone to failure; and that a shortage of skills to operate and manage sewered systems is an even greater threat to sustainability than funding gaps. It is generally the poorest and most vulnerable who wait the longest for service improvements, and who are most risk when services fail. There are compelling reasons to pursue less water-dependent, less-polluting approaches to sanitation improvement, but a comprehensive shift in approach is unlikely soon. In the interim, more robust systems are needed which have lower skills requirements and a reduced risk of failure. Given the urgency of service improvements – particularly in informal settlements – in a context of extremely high HIV and TB infection levels, South Africa does not have the luxury of plentiful time.

Keywords Water security and wastewater • Sanitation improvement • Sustainable sanitation • Urban sanitation • Water-borne sanitation • Subsidies • Service delivery • Wastewater management • AIDS • Accountability • MDGs • Rights based approach

---

K. Eales (✉)
Counterpoint Development cc, Pretoria, South Africa
e-mail: kea@iafrica.com

B. Schreiner and R. Hassan (eds.), *Transforming Water Management in South Africa: Designing and Implementing a New Policy Framework*, Global Issues in Water Policy 2, DOI 10.1007/978-90-481-9367-7_4, © Springer Science+Business Media B.V. 2011

4.1 Introduction

According to Paul Reiter, head of the International Water Association (IWA), 80% of future global water stress will come from population growth and development, not climate change (Reiter 2008). This perspective highlights growing water demand, particularly in a context of rapid urbanisation and the urbanisation of poverty. Implicitly, it flags the complementary growth in wastewater generation, and the need for effective wastewater treatment and water quality management on a large scale to minimise the contamination of water sources.

This chapter focuses on the linkages between water, sanitation and wastewater management in South Africa. It argues that the extremely high prevalence of HIV in South Africa adds particular urgency to ensuring that everyone can access clean drinking water and safe toilets in the short-term; and that in the longer term, raw water quality will determine South Africa's national water security as much as the quantities of water available. Effective management of wastewater is critical to both.

SA has set bold targets to ensure universal access to water and sanitation by 2014. Remarkable gains have been made since 1994: by April 2009, improved sanitation facilities had been provided to an additional 10.9 million people, equivalent to more than the total population of Portugal or Belgium. This is a remarkable achievement, by any measure. What is even more remarkable, given the significant poverty levels in South Africa, is that flush toilets had been provided to nearly nine million more people – nearly 20% of the population – since 1996, according to census data (StatsSA 2005, 2007). This means that nearly 60% of households nationally now have flush toilets.

But government's commitment to ensuring universal access to 'decent sanitation' does not yet align with the municipal capacity to support sewered sanitation as the default in all urban settlements, Moreover, extending coverage has been prioritised over sound maintenance of both new and existing infrastructure. Leaking networks, sewer spills and rising levels of wastewater treatment failure are polluting South Africa's water systems and, in time, could compromise national water security.

Wastewater management occurs at the interface between the separate policy frameworks for water resource management and water services; it remains comparatively marginal to both. Increasingly, South Africa's water security will be shaped by how effectively the country is able to manage its return flows and effluent streams – keeping the addition of nutrients, bacteriological agents and salts to a minimum, to protect living river systems and the services they provide, and to keep the cost of re-used water affordable.

This chapter maps some current challenges around service provision, sanitation improvement and wastewater management in South Africa. It argues that a rights-based approach to providing water and sanitation is a hollow promise if the service is not sustainable, and that a shortage of skills to operate and manage sewered municipal systems is an even greater threat to sustainability than funding.

Box 4.1 Access to water and sanitation in South Africa

Recent census statistics (StatsSA 2007) indicate that nearly half of all South Africans have an in-house water supply, while a further quarter have a water connection in their yard. Twenty percent source their water from a communal standpipe within 200 m, while around 4% do not yet have access to an improved supply. Fifty-six percent of the population has full water-borne sanitation. An estimated 8% have a ventilated improved pit toilet. One percent uses bucket toilets – a crude form of on-site toilet, which relies on a weekly municipal emptying service, and 27% make their own arrangements – usually a rudimentary pit toilet or open defecation (StatsSA 2007; DWAF 2008b) (Fig. 4.1).

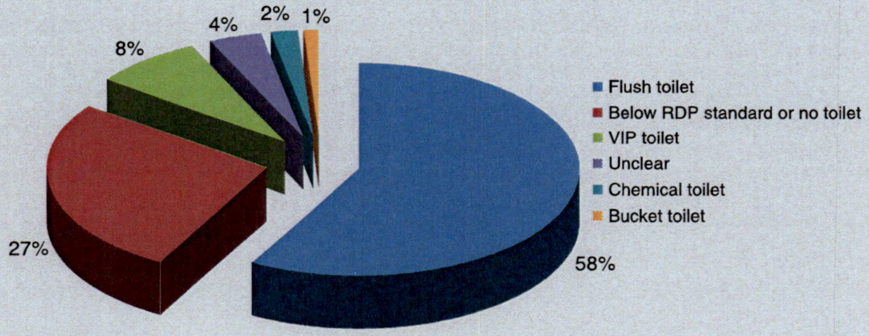

Fig. 4.1 Access to water and sanitation in South Africa, 2008 (Source: Derived from StatsSA 2007; DWAF 2008a)

4.1.1 Rethinking the Links Between Water and Sanitation

Sanitation is widely associated with technologies to manage human excreta, but the term has a far wider reach: it encompasses good personal and household hygiene and safe management of a range of waste streams, including wastewater and solid waste. Water for hand-washing and personal cleanliness is essential for good health and hygiene.

Faecal contamination of water can spread a wide range of diseases, from polio to those which have diarrhoea in common: cholera, typhoid, dysentery, giardia, shigella, and so on. It is ironic, then, that the default urban sanitation technology in the developed world systematically pollutes potable water with excreta, as water is used as a transport medium to convey excreta from households to central treatment facilities. This technology has high capital and operating costs, which together far outweigh the costs of a potable water supply; added to this are the growing resource costs of deliberately contaminating fresh water, and inadvertently contaminating aquifers and river systems with sewer seepage, spills and contaminated effluent from treatment failures. Given the spill-over consequences for water quality for people

Box 4.2 South Africa's rights-based approach to sanitation improvement

The African National Congress (ANC) came to power in 1994, with a mandate to utilise South Africa's immense resources to achieve social justice for all. The country's 1996 Constitution puts strong emphasis on building a culture of rights, and foregrounds human dignity as a founding premise of the new state. The Bill of Rights states that everyone has a right to dignity, a right of access to sufficient water, and a right to an environment that is not harmful to their health and well-being; it mandates the state to 'take reasonable legislative and other measures, within its available resources, to achieve the progressive realisation of each of these rights' (RSA 1996).

Sanitation features prominently in national political discourse and in the discourse of human rights. See, for example, this statement by the Minister of Water Affairs and Forestry in March 2007:

Since attaining democracy in 1994 and adopting our constitution in 1996, our government has an obligation to ensure that we promote and fulfil the rights of our people, importantly that includes the progressive provision of basic services such as water and sanitation. It is therefore fitting that we celebrate Human Rights Day in the same month that we have awareness campaigns such as the National Water Week and the Sanitation Week (Hendricks 2007).

living in unserved areas, conventional sewerage is regarded as profoundly anti-poor by a growing number of writers. (See, for example, Paterson et al. 2007).

Alternative technologies which do not pollute water are available to support safe management of human excreta, but most towns and cities have invested so heavily in sewered systems which combine human excreta with domestic wastewater, that there are substantial financial disincentives to pursue alternative technologies. This may change as a consequence of declining global reserves of mined rock phosphate. Phosphate is an essential nutrient for agriculture; the need to recover phosphate from human excreta to fuel agricultural production could spur major changes in sewage collection and treatment within the next 50 years.

4.2 The Water Resources Context

4.2.1 Growing Water Stress and the Importance of Good Wastewater Management

In a context of growing water scarcity, safe-guarding water quality is critical to ensure it remains fit for use. A key factor contributing to deteriorating raw water quality is the declining performance of many – if not most – municipal wastewater works.

As has been outlined in previous chapters, South Africa has low rainfall by international standards. FAO-Aquastat figures rank South Africa 11th from the bottom on an index of 50 countries in the world with the least annual renewable water availability per capita (UNESCO 2006), and the country is moving rapidly towards being classified as water scarce. South Africa has invested heavily in large dams and elaborate inter-basin transfer schemes to collect and convey water to where it is needed in the largely dry hinterland (Eales and Schreiner 2008). Population distribution tends to be concentrated in areas far from significant sources of water, and urban settlements are growing rapidly. With access to improved services and rising living standards, household water consumption is increasing, and urban water supply is the fasting growing sector of national water use (DWAF 2008a). This has implications both for water supply and wastewater management.

South Africa's rivers are fairly small by international standards, with limited flow volumes to mitigate contamination by the discharge of untreated effluent, particularly in the dry season. Thus South Africa has stringent effluent discharge standards to regulate return flows. Unlike many parts of Europe, Asia and South America which use rivers as an extension of their municipal sewer systems, South African municipalities must treat all municipal wastewater before it can be discharged. This adds significantly to the cost, infrastructure and personnel required to support waterborne sewage.

The volume of wastewater discharged to rivers nationally is rising rapidly, while the management of treated wastewater discharged by municipalities is generally poor. Just 3% of municipal wastewater treatment works nationally meet the requirements for a Green Drop certificate for wastewater management issued by the Department of Water Affairs (DWA 2010); this requires at least a 90% score against 11 weighted indicators of wastewater treatment and management. Even where the standard of effluent treatment is good, the sheer volume of contaminants and nutrients discharged is impairing river water quality. Over time, it is likely that even stricter wastewater quality management standards will be required, which will add further costs and complexities.

Declining raw water quality in many areas is raising new challenges around the availability of water and the fitness of water for use. 'Clean' water is needed to dilute contaminated return-flows and mitigate the effects of rising salinity and nutrient levels for agricultural and industrial users, and rising pollution levels mean increasing volumes of clean raw water must be diverted for blending, which adds to water stresses. The poor quality of discharged effluent is contributing to rising eutrophication and bacteriological contamination of rivers and dams. High nutrient loads stimulate the growth of cyanobacteria, which blocks sunlight in the water, which in turn prompts plant decay. The process of decay demands oxygen, and leads to visible fish kills and less visible ecosystem damage. Moreover, under the right conditions of pH and temperature, these cyanobacteria release toxins which can be harmful to humans if ingested – yet very few drinking water treatment works are equipped to remove these. Bacteriological contamination from poorly treated sewage can result in the spread of water-borne diseases. Those most at risk are people who drink untreated water, but where municipal water treatment is inadequate, even those drinking tap water are at risk.

4.3 The Municipal Services Context

There has been a massive expansion in water and sanitation infrastructure coverage in South Africa since the country's first democratic election in 1994. In May 2008, South Africa's Minister of Water Affairs and Forestry Lindiwe Hendricks told Parliament that the country had now met its MDG goals for both water and sanitation (Hendricks 2008a), having met the water targets 7 years previously against a 1994 baseline.

By 2014, government aims to provide universal access to water and sanitation; this surpasses the MDG target of halving the proportion of those without access by 2015. Government, moreover, is funding the full cost of providing taps and toilets in low income areas. In urban areas, there is strong commitment to meeting the sanitation target with flush toilets despite the range of alternative sanitation technologies available. Given the realities of South Africa's growing water scarcity and the rising direct and indirect costs of treating effluent, alternative technologies warrant serious consideration. But South Africa's emphasis on providing flush toilets throughout its urban areas has to be understood in the context of the country's history of apartheid, under-development and statutory racial discrimination. Flush toilets are a powerful symbol of dignity, modernity and aspiration to a better life, and anything other than a flush toilet is regarded as second-best, discriminatory, and at best an interim option until 'a proper toilet' can be installed.

With hindsight, it is evident that government underestimated the intensive resource and skills requirements of conventional reticulated systems. Flush toilets need house connections for water, and result in increased water consumption which may call for new water resource and bulk infrastructure development. Most of the network is buried and out of sight, and the extent and complexity of the water and sewer network is often not appreciated. Most water leaks and sewage spills are underground, and require specialised monitoring and maintenance equipment. Managing the biochemistry of safe water treatment and effluent management requires high levels of expertise, as well as ongoing laboratory testing to ensure compliance with health, safety and environmental standards. Quick response times are needed in case of bursts, spills and outages to minimise the risks to public health and the natural resource base. And failures, which are inevitable, impact most harshly on the poor as they tend to have the least clout in getting their problems heard and fixed, and the most meagre resources to cope with the consequences of contaminated drinking water, faulty services, leaks and spills.

In its commitment to providing high level services, South Africa is at risk of developing service infrastructure which it does not have the financial and human resources to operate sustainably, and which has a high risk of failure with consequences for water quality and human health. Notwithstanding the ANC government's commitment to providing 'safe water and decent sanitation to all,' the real issue is the reliability and sustainability of those services. Failures impact not just on the quality of services provided to local users; sewage spills compromise the health of South Africa's rivers and aquifers, and potentially the health of downstream users too.

4.3.1 Wastewater Treatment Failure and Service Expansion

One of the most significant causes of deteriorating water quality in South Africa is the discharge of poorly treated municipal sewage. Addressing Parliament on 11 March 2008, Minister Lindiwe Hendricks said,

> We have done an audit [of municipal wastewater treatment works] and found that the situation in many municipalities is dire, and must be addressed as a matter of urgency. The pollution in some of our rivers can be directly linked to failure on the part of these municipal waste water treatment plants, and there is no denying that some of these plants are in poor condition (Hendricks 2008b).

The immediate reasons for the poor state of municipal effluent discharge in many areas were revealed in a 2006 survey conducted for government, which investigated a representative sample of 51 plants in eight provinces. The survey found a critical shortage of trained and skilled staff, particularly experienced process controllers and mechanical/electrical maintenance staff. It reported that 56% of the plants lacked the skilled staff to maintain the installed mechanical/electrical equipment and instrument adequately, while 50% were understaffed and needed additional skilled operators (Snyman 2007). Just 4% of the plants surveyed were operated and maintained adequately, and 'immediate intervention" was needed in 30% of works to avoid health crises. Even fairly low-tech waste stabilisation, or oxidation, pond systems are suffering from accumulated neglect and poor operation. A separate 2006 survey of 47 Free State oxidation ponds revealed that over half were illegally discharging effluent with high faecal coliform loads into local rivers or streams (Van Vuuren 2006).

There is a link between poor effluent treatment and poor drinking water quality where water treatment processes are inadequate. The risks of malfunctioning treatment and poor water quality are highest in the smaller centres, through a combination of understaffing, underfunding, erratic chemical dosing, technical failure and poor monitoring (Momba et al. 2006, 2008). A recent sample of 53 small water treatment plants in the Eastern Cape showed that just 34% of plants complied with the mandatory national standard for faecal coliforms, while an even lower percentage – 29% – complied at the end of the distribution network, at point of use (Momba et al. 2008). These figures are cause for grave concern, as the Eastern Cape is the second poorest province in the country, and it is poor households who suffer the harshest consequences of unsafe drinking water. There have been a number of severe diarrhoea outbreaks where faecal contamination of water has not been neutralised through effective water treatment, and several instances of typhoid – most notably in Delmas, in Mpumalanga, where at least 13 people died in 2005 (NISC 2005–2008). The deaths of more than 140 infants over a 3 month period in the Ukhahlamba district in early 2008 prompted sober reflection on the extent and causes of current treatment failures (DOH 2008; ECDOH 2008; Ukhahlamba District Municipality 2008). A significant contributor to these failures is the rapid pace of new infrastructure development, in a context of far-reaching municipal restructuring and staff changes. Since 1994, nearly 2.5 million houses have been built and transferred, free, to poor households in urban areas. With few exceptions, these houses have flush toilets. In addition, there has been a national programme to eradicate bucket toilets. Since 1994, nearly 500,000 bucket toilets have been replaced, mostly with reticulated flush toilets (Office of the Presidency 2008), and half have been replaced in the last 3 years (DPLG 2008) See Endnotes.

In South Africa, flush toilets have historically been associated with white privilege, and dry toilets with discrimination. The dormitory black townships of former 'white' South Africa generally had a lower grade version of the reticulation networks servicing 'white' areas, with bucket toilets and pit toilets servicing those not on the network. Post-1994, there has been little debate about the merits of extending waterborne sanitation, notwithstanding the poor condition of many existing networks, and the sheer magnitude of urban sanitation backlogs. Attempts to implement lower cost permutations of wet systems, such as non-sewered flush toilets with an on-site soakpit, failed in new settlements such as Ivory Park, outside Johannesburg, in the 1980s, primarily because of poor execution, and pilot projects to introduce condominial sewering in Johannesburg and Durban have not been scaled up or replicated, because residents regard them as a lower standard alternative, and are not willing to take on responsibility for local maintenance when adjacent settlements with conventional sewering are not required to. Thus the default service level in formal settlements is fully reticulated flush sanitation.

South Africa's national sanitation policy (RSA 2001) speaks primarily to basic sanitation in a context where rural households use on-site dry toilets. It is silent on what sanitation improvement means in a context of urban reticulated services, and says almost nothing about grey water and wastewater management. Urban sanitation improvement focuses overwhelmingly on providing or upgrading toilets. There is no question that investment in infrastructure and effective service provision is an essential component of sanitation improvement in dense urban settlements, but sanitation improvement requires more than the delivery of infrastructure and access to toilets. It requires awareness of the linkages between water, hygiene and health improvement – which is critical given high HIV and TB prevalence. It calls for practical guidance around grey water management, and the importance of safe waste disposal, together with information on how poor solid waste management not only compromises public health, but can also clog sewers and storm water systems and cause blockages and spills. But because of the emphasis on infrastructure and service provision, sanitation improvement has become something that government delivers, rather than a public-private partnership in the most literal sense. The emphasis has been on toilets, not people.

Expansion of water and sanitation coverage since 1994 has been characterised by a tension between the urgent need to improve access to water and sanitation and the equally urgent need to build the institutional capacity to operate and maintain those services. The ability to source funds for new infrastructure has frequently outpaced the development of the institutional capability to operate and manage that infrastructure.

Moreover, the focus of infrastructure improvement has been on the 'user interface' – taps and toilets – rather than the broader infrastructure package and institutional arrangements needed to operate and maintain the system effectively. One inevitable result is significant bulk infrastructure bottlenecks. This is particularly evident in settlements where bucket toilets have been removed in an accelerated eradication programme since 2005.

4.3.2 *The Bucket Eradication Programme in the Free State*

Spurred partly by the MDGs, South Africa's 2003 *Strategic Framework for Water Services* set 2006 as the target date by when all bucket toilets would be removed. The definition of the target was subsequently narrowed to refer to bucket toilets in formal settlements established before 1994, and the deadline was extended to December 2007. Bucket eradication was defined as a national priority, and government allocated grant funding of ZAR1.8-billion (US$22 million) to achieve the target of removing roughly 250,000 bucket toilets (excluding a comparable number falling outside the revised definition) (DPLG 2008). Addressing a Round Table Discussion on Accelerating Infrastructure Delivery in March 2007, the Minister of Provincial and Local Government told over 500 mayors, government officials and senior political leaders that '[t]argets that have been set are not open for review and no Municipality will be exempted from meeting them' (DPLG 2007). Municipal performance was monitored closely and immense pressure was applied to ensure they met the target.

The greatest number of bucket toilets was in the Free State province. Between February 2005 and July 2008 the province reduced the number of targeted bucket toilets by more than 90%, from 127,658 to 12,572 (DPLG 2008). The vast majority were replaced with flush toilets, as residents rejected VIPs vehemently. A range of challenges were highlighted in a 2006 DWAF planning report (DWAF 2006), in the early stages of the Free State bucket eradication programme:

- Overall water consumption and wastewater generation would rise sharply in most municipalities, because of water use for flushing, and the associated on-site or in-house taps provided in some settlements for the first time.
- Most towns did not have local water resource development potential and would have to import water from distant water sources, requiring regional bulk water schemes.
- Thirty-five percent of water treatment works required upgrading to accommodate increased water demand associated with new flush toilet connections – some water treatment works were over 50 years old.
- Few of the existing wastewater treatment works had adequate spare capacity to accommodate increased sewage inflows. Forty-one percent (33 out of 80) were operating over-capacity already, and there was significant non-compliance with wastewater effluent quality requirements. 45% were failing on bacteriological indicators, 45% were not complying with COD standards, and 55% did not comply with ammonia standards. The main reasons cited for non-compliance

included operational inefficiency, equipment shortcomings, lack of skills, inadequate monitoring equipment, and lack of proper chemical dosing equipment.
- Fifty-five of sewage treatment works needed to be upgraded to accommodate increased wastewater production
- Skills deficits to operate and maintain infrastructure
- Concerns about the additional financial burden on municipalities due to their increased service responsibilities, a limited economic base, high rates of poverty and unemployment poor cost recovery and aging infrastructure.

The DWAF report estimated that in the Free State, a budget of just under ZAR1.5 billion (US$187.5 million) would be needed to implement the programme, excluding water resource development requirements (DWAF 2006). Funds allocated by government and from municipalities' own sources fell substantially short of that amount, and spending to date has been even less. Subsequent unit costs ranged from ZAR5,905 (US$738) to ZAR32,306 (US$4 308), but it was investment in bulk infrastructure rehabilitation and upgrading that was the main casualty of the funding shortfall. Despite evidence of effluent overflowing from both oxidation ponds and conventional works in several towns (Mafereka 2007), 110,000 buckets had been replaced with new flush toilet connections by July 2008 (DPLG 2008). Water supply shortages were remedied in at least six municipalities by using drought relief funds from government to drill new boreholes (National Treasury 2008a; De Kock 2008).

Data presented at a Free State water quality management meeting convened by regional DWAF office in July 2008 suggests that wastewater treatment failure has risen sharply in the province since 2006. In February 2008, treated effluent failed to comply with the required standards in 52 out of 81 towns (64%), with just seven recording few or no failures (DWAF 2008d). By September 2008, non-compliance had risen to 74% (Fig. 4.2).

DWAF is working closely with Free State municipal councillors and officials to identify the causes of the problems and develop remedial strategies to fill posts, source funds for maintenance and repairs and improve local monitoring and responsiveness. The department is putting emphasis on raising awareness, particularly among municipal councillors, about why good effluent management matters, and has developed a simple 'dashboard indicator' to show compliance performance and promote greater accountability (De Kock 2008). There is evidence of substantial progress in strengthening water quality management.

But there are two major challenges facing the Free State, and South Africa as a whole, which stack the odds against successful remediation of the growing wastewater management crisis for as long as we continue down the current path: money and skills.

4.3.3 Funding and Affordability

In aggregate terms, South Africa is classified as a middle income country, with a GDP per capita (PPP) of US$11,100 in 2005 (UNDP 2008). This average masks

Fig. 4.2 Wastewater compliance in the Free State Province, February 2008 (Source: DWAF 2008c)

extremes of wealth and poverty, worsening income inequality and one of the highest Gini co-efficients in the world. Nonetheless, one consequence of South Africa's wealth and relatively advanced economic development is that the country has the resources to fund bold development programmes to remedy distorted access to water and sanitation. Significant amounts of funding are being transferred to municipalities to address infrastructure backlogs and fund service provision to poor households.

Beyond municipalities' own funding sources, the two most significant grants are the Equitable Share, which is intended to cover the costs of providing basic services to poor households; and the Municipal Infrastructure Grant, or MIG, which is intended to cover the capital costs of providing infrastructure for basic services for poor households. MIG funds enable the rapid development of infrastructure, but also incentivise swift measurable spending, rather than enduring outcomes. When money is free, considerations of cost-effectiveness and life-cycle costing and sustainability requirements are not necessarily given the primacy they warrant. Even more worrying is growing evidence that the MIG programme is funding the construction of water services infrastructure in many areas which requires greater subsidies than are currently available, and operating and managerial skills that are in short supply. The more the poorer municipalities extend infrastructure coverage to service poor people, the more their revenue needs rise, and their financial viability is likely to deteriorate unless there is a decisive shift to more pragmatic service levels, or unless government steps in to fill the breach by raising subsidies for operating costs further. But there is little assurance that increased funding will resolve the problem in the absence of effective municipal planning, technical and financial management systems and greater public accountability. Moreover, it would require extremely skilled political leadership and a very different style of public engagement to persuade people to accept lower service levels than those in adjacent settlements.

The stark reality is that over 40% of South Africans live in poverty, and at present only about 43% of the working age population is working. In countries in Latin America, Eastern Europe and East Asia at similar levels of development, the proportion is about 50% higher (Hausman 2008). Far-ranging economic restructuring has seen huge job losses in the three sectors – agriculture, mining and manufacturing – which traditionally employed unskilled workers, and the high growth sectors are mostly skills-intensive. The consequence is widespread joblessness. Welfare support through social grants has blunted some of the most acute forms of poverty over the past 5 years, but a significant and growing proportion of South Africans cannot afford to pay for the services they receive.

Municipal revenue is rising less rapidly than the growing cost of providing services as coverage is extended (Savage et al. 2008). One consequence is growing municipal dependence on grant funding for operating expenditure. Dependence on grants for municipal revenue requirements has risen from 16.1% in 2003/04 to 25% in 2007/2008, and is projected to reach 31.9% by 2009/2010; this is most marked in the six big metros, where dependence has risen from 4.6% of total revenue to 16.9% in 2007/2008, and is projected to reach 20.3% by 2009/2010 (National Treasury 2008b).

Further development of high level infrastructure and growing reliance on grants presupposes that South Africa will at least maintain its current economic performance and will be able to afford ongoing transfers on a large scale for the foreseeable future. This may well be achievable, but it is not without its risks.

An important consequence of this grant dependence is that it shifts the emphasis of municipal accountability away from its primary constituency, local residents and enterprises, to its relationship with national government (Savage et al. 2008). Local accountability is weakened. And what national government currently holds municipalities accountable for is their performance in spending funds and delivering quantifiable outputs – number of buckets removed, number of toilets delivered, and so on – rather than infrastructure quality and service performance. Cost-efficiency is not a primary concern for municipalities when the emphasis is on absorbing large tranches of funding to avoid being reprimanded for under spending or penalised with reduced allocations next year.

Growing municipal debt is a further consequence of providing under-funded services to poor people. By December 2007, consumer debt to municipalities stood at ZAR44.1 billion (US$5.1 bn) (National Treasury 2008b). Financial administration is weak in the majority of municipalities, and there are enormous political sensitivities around taking action against poor debtors. In many areas, the gap between expenditure and income is being funded through under-spending on maintenance and inadequate provision for rehabilitation and renewal, leading to infrastructure decay. Dry taps, pump failures, pipe bursts, spills, leaks and declining water and effluent quality all signal the need for urgent interventions, rehabilitation and renewal. Increasingly, municipal maintenance is reactive (Lawless 2007). This problem is particularly acute in the still predominantly black townships, because of their history of neglect and under-development, but leaks, bursts and sewer spills are becoming more frequent in all areas.

But municipal neglect of spending on maintenance, renewals and upgrades is not only a consequence of a funding squeeze; if it were, it could be remedied to some extent through further grant funding, at least in the short term. The far larger challenge is the skills gap in many municipalities.

4.3.4 Staffing and Skills for Municipal Service Provision

There has been a significant loss of institutional memory and technical expertise in most municipalities since 1994 which has undermined the development, operation and maintenance of water and sanitation services. Between 1994 and 2003, municipalities nationally went through a process of almost continuous restructuring and transformation – first to amalgamate and integrate previously race-based administration; then to bring all areas, including the former homelands, under newly demarcated municipal administrations; and then to assign powers and functions across metropolitan, district and local municipalities and build the capacity to execute them. All of this was accompanied by far-reaching staff changes, as personnel were 'rationalised,' transferred, recruited, promoted or demoted. Pressure to meet employment equity targets added a further dimension, as municipalities sought to make appointments that reflected better the demography of the country. But the pool of available technical and managerial staff available in South Africa to power service delivery does not yet align with employment equity targets, and many municipalities have not filled key positions.

The enormous expansion in networked services over the past decade (see Fig. 4.3) has not been matched by similar expansions in staffing and expertise. As described in Chapter 3, municipalities now employ far fewer engineering professionals than they did 15 years ago, despite an exponential increase in service infrastructure. These shortages impact across the water services delivery chain and into water resource management, and place immense strain on existing municipal technical staff. High turnover in professional staff erodes the institutional memory needed for durable service delivery. Staffing shortages impact particularly heavily on planning, design and project management. Reforms in artisanal training have led to a severe shortage of key skills needed for a range of operational and maintenance functions.

Reticulated water and sanitation connections require intensive technical support capability; but in the same period that installation of new taps and flush toilets soared by over 80%, the number of municipal engineering professionals per capita nationally plummeted by 80%.

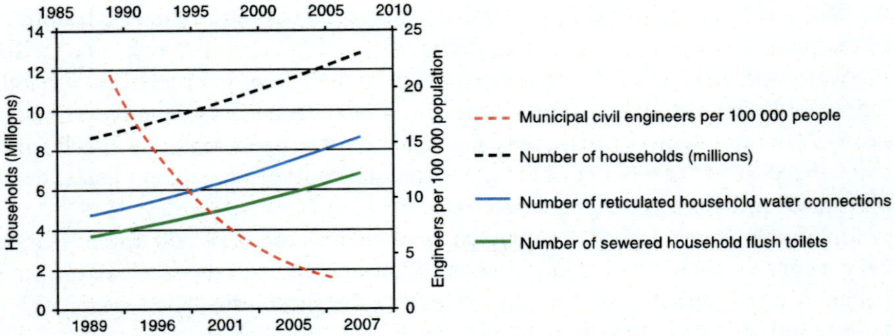

Fig. 4.3 Funding transfers to municipalities, by type, 2003/04–2009/10 (Source: National Treasury)

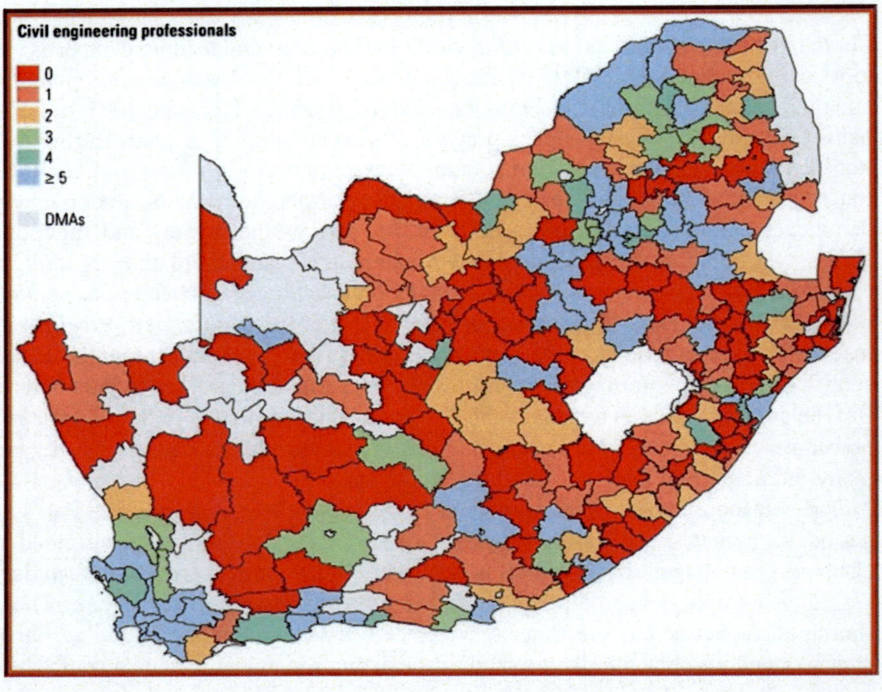

Fig. 4.4 Key trends in networked water and sanitation, 1989–2008 (Source: Derived from Lawless, 2007; StatsSA, 2001, 2007; DWAF 2008)

Figure 4.4 illustrates the findings of a national survey of technical capacity across all services in municipalities in 2005. The darker the shading, the more severe the skills gap, with red-shaded municipalities having no civil engineering professionals at all. The 2005 survey found that 83 municipalities had no civil engineering staff at all. The situation has deteriorated since this survey was done (Lawless 2007, 2008).

Engineers are now being seconded into local government through the Siyenza Manje programme, a joint initiative of government and the Development Bank of South Africa, and an innovative mentoring programme called ENERGYS (Engineers Now Ensuring Rollout by Growing Young Skills) is assigning experienced engineers to work in municipalities alongside recent engineering graduates to help build their skills base through structured experiential learning programmes (Lawless 2007). There is growing recognition that the sector must grow its own pool of expertise to meet the sectors' demand for skilled professionals, artisans and practitioners. But it will be many years before new graduates acquire the experience needed to lead effective municipal service delivery.

The evidence suggests that South Africa is investing in new infrastructure which requires operating and maintenance skills that do not align with the skills available. Perhaps the most significant constraint on sustainability is the managerial and technical expertise available within municipalities to plan and manage vastly expanded service provision.

4.4 Municipal Delivery Approaches

Municipalities interpret their responsibility for service delivery very literally, and very few seek to explore formal service partnerships with community-based organisations, let alone the private sector. There is significant opposition to anything resembling an abdication of responsibility by the public sector, and strong rejection of any form of privatisation in the water sector, in part because of a desire to retain and maximise public sector employment opportunities in a context of chronic unemployment. The sale of assets involved in the delivery of basic services is specifically prohibited by law. There has been no privatisation of the management of water services beyond a few isolated instances in the 1980s and 1990s; but there is widespread dependence on engineering companies to provide extensive support.

One consequence of the target-driven approach to service improvements is that public involvement in planning and achieving service improvements has declined significantly. Equipping local residents to participate meaningfully in development planning, decision-making, implementation and oversight takes time, and with large budgets to spend and daunting targets to meet in tight timeframes, few municipalities have the resources to accommodate meaningful public engagement. The result is that rapid supply-side contractor-driven delivery becomes an end in itself, even if the quality and performance of the new infrastructure and services falls short of expectations.

Because government has set itself up as the sole driver of service improvements, and provides immense funding and resources to municipalities to deliver fully-funded infrastructure, there are few incentives and little support for independent self-help initiatives by citizens living in areas that are not yet on a municipal project schedule. Inevitably, there is rising impatience and frustration with inadequate services. Political leaders generally respond by calling for patience, co-operation and understanding until improvements can be made, with a number of unfortunate consequences.

On the one hand, this approach entrenches a culture of passivity and dependence on government for service improvements, and growing resentment and frustration. There have been periodic explosions of protest and public violence – including the horrifying attacks on foreign nationals living in informal settlements in May 2008, which were stoked at least in part by the stress of sharing inadequate services. Where there are already 50 or more people sharing a single toilet, each new resident adds significant stress, and outsiders are scapegoated.

On the other hand, it adds to the pressure to accelerate the pace of delivery, which in turn accentuates supply-driven, top-down infrastructure planning and delivery in ways that marginalise the intended beneficiaries and raise the risks of service failures.

4.5 Towards a New Approach

4.5.1 Some Drivers for a New Approach

South Africa's population has doubled since the mid-1970s, and has grown by over 20% since 1996. Moreover, average household size has fallen, which means that the number of households who need services has grown from approximately 9 million in 1996 to 12.5 million in 2008. This translates to a 39% increase, on top of existing service shortfalls and backlogs. In Gauteng province, home to Johannesburg and two other metropoles, the number of households has grown by 56% (Office of the Presidency 2008). This is putting immense strain on existing infrastructure, and enormous pressure on municipalities struggling to eradicate housing and service backlogs in a context of rapid household increase.

There has also been a significant increase in informal settlement since the mid-1980s, when urban influx controls were lifted. Informal settlement includes both backyard shacks, where tenants share the services provided to the main house, and free-standing informal settlements on the edges of even the smallest towns, where service inadequacies are most acute. Government aims to eradicate all informal settlements by 2014 and, wherever possible, relocate residents to new houses with full services in formal settlements. But with an urban housing backlog of at least 2.2 million units, it is unlikely that this target will be met. In-situ upgrading makes the most sense, but this is not feasible in areas where people have settled in floodplains or on severely dolomitic land. Resettling people elsewhere calls for alternative land parcels, but available public land close to the urban core has mostly been sold off to developers, which means new settlements are generally located far out of town. This perpetuates spatial apartheid and urban sprawl, and raises the cost of living – and particularly transport – for poor households. There is also evidence that a number of re-housed people are renting or selling their new houses and moving back to informal settlements. Some say they cannot afford the high costs of servicing, while others point to high transport costs (Hetherington 2008; FinMark Trust 2006).

Services in many informal settlements are severely inadequate. Some municipalities provide communal toilets, but maintenance problems, vandalism and the personal safety of users are endemic challenges. Some provide serviced chemical toilets, but their operating costs are relatively high, and the service intervals and number of facilities provided generally fall far short of what is needed. Some municipalities continue to issue buckets and provide a bucket service. VIP toilets are being built in some settlements where there is no imminent prospect of in-situ upgrading or relocation, but the density of many settlements means that these are often built around the periphery of the settlement and offer only a partial solution, particularly to women and children needing a toilet after dark in a context of high crime. In addition, residents often reject VIP toilets, partly because they would prefer flush toilets, and partly because provision of VIP toilets is taken as proof that they will not be rehoused any time soon. All of these considerations have implications for water quality from unmanaged run-off.

Improving sanitation and environmental health in informal settlements is an urgent priority for a range of reasons, but perhaps the most pressing is that the prevalence of HIV and AIDS is highest in informal settlements. National HIV prevalence among adult South Africans is now estimated at 19%. One recent study noted that while the prevalence of HIV nationally among women aged 25–29 is now one in three, this figure is even higher in informal settlements (CADRE 2008). HIV, moreover, is now widely associated with TB, and South Africa now has the highest incidence of TB in the world, at 998 per 100,000 people (WHO 2008).

HIV, AIDS and TB have profound implications for water and sanitation services. Unhealthy living environments heighten the risk of illness, and HIV positive people are particularly prone to diarrhoea, and need close access to sanitation facilities and safe water. Care-givers need additional water to cope with increased laundry and to support good hygiene and care. And households affected by AIDS often have reduced incomes and additional expenses, with profound implications for their ability to pay for services.

There can be little question that South Africa urgently needs to find low cost, low skills ways of providing 'decent sanitation' through approaches which have a reduced risk of failure. Moreover, adaptation to climate change will require low energy technologies, and, increasingly, approaches which minimise the risk of water contamination.

4.5.2 Two Promising Wastewater Treatment Technologies

The majority of South Africans are unlikely to deviate from their insistence on flush toilets in the foreseeable future, and it is likely that any change in government's current target-driven top-down approach to service delivery and sanitation improvement will require several years of lobbying and debate. In the interim, two wastewater treatment systems offer immense promise as a response to some of the current challenges. Both are low cost, have low skills and energy requirements, are robust and have a low risk of failure.

4.5.2.1 DEWATS

DEWATS, or Decentralised Wastewater Treatment Systems, provide a modular technology for treating up to 1 megalitre of organic wastewater flows daily with the core of the system being an anaerobic baffled reactor. Over 500 DEWATS units have been installed in Asia over the past decade, primarily in India and Indonesia, and are used to serve dense urban settlements, hospitals and agricultural industries. For neighbourhood-level sanitation systems, DEWATS can be used as a shared facility – like a large septic tank – linking anything between 10 and 60 households in poor and middle-income areas, often in combination with simplified sewers; or as the wastewater treatment facility for a community sanitation centre offering toilets and washing facilities (BORDA 2005, Ismawati 2007; Ulrich 2008).

Construction of the facility is comparatively inexpensive, and responsibility for operation and maintenance can be assigned to a local resident who receives training and is employed by a committee representing users. Users pay a small monthly fee for use and to cover maintenance costs, and the system is then financially self-sufficient (Ulrich 2008; Ismawati 2007). A significant benefit of this approach is that local users hold each other accountable for payments, responsible use and good maintenance, and are not dependent on a remote agency managing a costly, complex technology with onerous maintenance needs, over which they have no control.

DEWATS reduces the Chemical Oxygen Demand (COD) of effluent by up to 90%, but it does not reduce nutrient or pathogen levels. Disinfection is an easy add-on, but high residual Nitrogen and Phosphorus levels are more challenging, and preclude discharge to open water bodies. Instead, nutrient-rich effluent can be used readily to irrigate food and non-food crops. DEWATS holds enormous promise for South Africa, particularly in informal settlements where there is an urgent need for improved communal facilities – certainly until such time as the settlement is upgraded, or until residents are relocated to a new settlement. However, any technology evolves in a particular social context, and there are big risks in implementing a given technology in a very different context, with different social dynamics and accountability relationships. Moreover, DEWATS was developed to serve ‘washers’ who use pour-flush toilets, and water for anal cleansing. One of the first challenges in adapting this system for use locally is that the vast majority of South Africans are ‘wipers.’ Those who can afford to, use toilet paper, but newspaper is used widely; this significantly raises the risk of blockages, particularly in communal toilets where accountability for responsible use is more diffuse. These challenges can, nonetheless, be overcome.

A key benefit of DEWATS in South Africa is that it would make flush sanitation feasible far sooner in urban settlements that are far from existing sewer networks. Particularly for people living in poor conditions in informal settlements, this is very good news.

4.5.2.2 Integrated Algal Ponding Systems

IAPS, or Integrated Algal Ponding Systems, are a significant advance on conventional pond systems, and utilise both aerobic and anaerobic processes. The daily treatment capacity of IAPS is limited primarily by the space requirements of the pond system, and 2 megalitres a day is generally the limit. An experimental IAPS plant has been run at scale in South Africa successfully for over 10 years, and its wider application is now being investigated carefully, both as a stand-alone system and to provide supplementary disinfection and nutrient removal capacity to boost the performance of failing treatment works currently contaminating local rivers (EBRU n.d.; Timm 2008).

Final effluent from the IAPS complies with all DWA requirements for discharge to an open water course except for Chemical Oxygen Demand (COD), which is at around 100 mg/l. This is mainly due to the residual algal component that is not removed in the settling pond. This residual provides an oxygen-contributing rather than an oxygen-demanding component to the final stream, and arguably should be exempted for this reason. In an open stream, the algae would not consume oxygen, but would rather add to the supply of dissolved oxygen (Munnik 2008).

Several municipalities are now considering installing IAPS plants. The Amathole District Municipality, for example, is assessing whether IAPS plants could replace existing conventional treatment plants in several small towns which need extensive rehabilitation (Timm 2008). Given the large number of small, understaffed, under-resourced treatment works around the country in urgent need of repair, upgrading or rehabilitation, wider use of IAPS plants offers a durable, low cost remedy to one dimension of the challenge of providing flush toilets as the default.

4.5.2.3 Limitations and Opportunities

These systems are not cure-alls, and they are not a substitute for large-scale urban treatment works capable of treating large volumes of water swiftly within compact footprint. Key advantages of both the DEWATS and IAPS systems are that they can be operated by relatively unskilled people, are extremely robust and have a low risk of failure. Capital and operating costs are significantly lower than conventional treatment works, there are few moving parts, no chemical additives are needed, and the electrical energy requirements are nominal. Neither system is vulnerable to power outages – unlike conventional systems, where power cuts can lead to severe process failures.

4.6 Conclusion

The sector slogan 'Sanitation is Dignity' has deep resonance in South Africa, given the country's history of statutory discrimination and grossly inequitable service provision. The importance of affirming people's right to dignity is not negotiable, but the emphasis on providing flush toilets as the default in urban areas is inhibiting more comprehensive sanitation improvements which engage hygiene practices, and is leading to investment in infrastructure that has dubious sustainability. It is generally the poorest and most vulnerable who wait the longest for service improvements, and who are most risk when services fail. Extremely high HIV and TB prevalence rates, particularly in informal settlements, underline the urgency and importance of achieving sustained improvements in sanitation nationally.

The volume of wastewater discharged to South Africa's rivers nationally is rising rapidly, while its quality is deteriorating in many areas. The state of sewage treatment in many municipalities is indeed 'dire,' as the Minister of Water Affairs has noted, because South Africa relies heavily on treatment technologies with high skills requirements, and because the costs and maintenance requirements of networked services have been underestimated. South Africa urgently needs to find low cost, low skills ways of providing 'decent sanitation' through approaches which have a reduced risk of failure. In the context of flush toilets, two immediate possibilities are decentralised wastewater treatment systems, and integrated algal ponding systems.

There is, however, a far larger and more important dimension to service improvements: the nature of the relationship between citizens and the government. As noted in Chapter 3, the participation of citizens in planning and overseeing seeing service delivery has been subordinated to top-down service delivery, and citizens have, to a large extent, been left on the sidelines of the local government system. Intent on delivering services, municipalities have forged ahead, with inadequate engagement with citizens in jointly determining needs, priorities and management approaches. Many citizens understandably have responded to government's commitments to deliver tangible improvements to their quality of life with high expectations and rising impatience. This, in turn, is fuelling fast-tracked top-down delivery approaches and widespread service failures.

A large-scale shift away from costly, risky, resource-intensive waterborne systems is unlikely in the near future. Any shift would need to target upper-income households, because of the current association between privilege and flush toilets; the drivers for a comprehensive shift – primarily water, energy and phosphate constraints – are still evolving. Significant interim gains towards enhanced sustainability and a reduced risk of failure can, nonetheless, be made by utilising wastewater treatment technologies suited to small towns and settlements which are currently struggling to align the technologies they have with the skills and funding available to run them.

But lasting sanitation improvement requires more than investment in infrastructure. Good infrastructure is necessary to collect and treat human excreta, but it is not sufficient to achieve good sanitation, which requires an active linkage between personal and household hygiene practices and effective management of excreta, wastewater and solid waste. Government can and does provide sanitation infrastructure, but it cannot deliver good hygiene practices, and good hygiene is one of the most effective ways that people with compromised immunity can defend their health. Sanitation improvement in South Africa needs to reach beyond delivery of infrastructure to meet service delivery targets. If it does not, the outcomes could compromise the services needed to support good personal and public health, and South Africa's national water security.

4.7 Endnotes

In Adelaide and Bedford, two small towns in the Eastern Cape, and perhaps elsewhere, residents who received flush toilets as replacements for the bucket system have now reverted to using bucket toilets again, because the municipality cannot provide water for flushing and the sewage treatment works cannot cope with the increased volumes; residents complain that the revived bucket service is worse than it was before (Timm 2008). Some Gauteng residents have insisted on keeping their buckets and bucket toilet structures, to serve their backyard tenants (Omigeer 2007). Some residents use their old buckets to pour-flush their new flush toilets, because bad workmanship and cheap fittings means their plumbing leaks (Ndlovu 2007). And the Premier of the North West Province noted in her State of the Province speech in March 2008 that several municipalities which had succeeded in eradicating buckets were now re-issuing them to new residents:

> In terms of the Presidential targets for bucket removal, the province has successfully eradicated all bucket toilets that were identified as 1994 backlogs. Sadly, we all know that a task of eradicating buckets, such as that of eliminating shacks, is an endlessly moving target. For every shack removed, another one or more mushroom elsewhere. For instance, by the end of November 2007, 16 500 buckets were eradicated across NMM, BDM, Maquassi Hills and Matlosana Municipalities. And yet another large number has mushroomed in the same municipalities. We have moved over many hurdles and battled the hard and unfriendly terrains of underdevelopment, in order to ensure that we serve our people to the best of our ability. Even this commitment is not sufficient (Molewa 2008).

References

BORDA (Bremen Overseas Research and Development Association) (2005) DEWATS. Decentralised wastewater treatment systems. Retrieved November 25, 2008 from http://www.borda.de/conrat/Dewats_Mai05_72.pdf

BORDA South Asia (2008) Operational tasks for the upkeep of Decentralised Wastewater Treatment system (DEWATS). Retrieved December 10, 2008 from http://www.borda-sa.org/

CADRE (Centre for AIDS, Development, Research and Evaluation) (2008) Talking about AIDS in an informal settlement. Experiences from Orange Farm, Gauteng. Retrieved December 12, 2008 from http://www.cadre.org.za/files/CADRE percent20Orange percent20Farm percent20Report percent20Final.pdf

De Kock C (2008) Personal communication, Bloemfontein, 13 August 2008

DOH (Department of Health) (2008, April) Report on childhood deaths, UKhahlamba district, Eastern Cape. Retrieved December12, 2008 from http://www.health-e.org.za/documents/7f3d48106e60e6e5d164698624c5efb3.pdf?PHPSESSID=1a9e0a0951

DPLG (Department of Provincial & Local Government) (2007, November) Round table discussion: government targets are not negotiable. (Record of discussion between the Minister of Provincial and Local Government and 500 national, provincial and municipal delegates, Midrand). Retrieved November 17, 2007 from http://www.thedplg.gov.za/index.php?option=com_content&task=view&id=381&Itemid=48

DPLG (Department of Provincial & Local Government) (2008, August). Quarterly performance of the Municipal Infrastructure Grant (MIG) for 2007/2008. (Presentation to the Select Committee on Finance. Department of Provincial and Local Government). Retrieved December 19, 2008 from http://www.pmg.org.za/report/20080814-quarterly-performances-provinces-and-municipalities-conditional-grant

DWAF (Department of Water Affairs & Forestry) (2006) Assessment of financial viability of water services authorities. Department of Water Affairs & Forestry, Pretoria

DWAF (Department of Water Affairs & Forestry) (2008a). Water for growth and development strategy. Draft Discussion Document, May 2008

DWAF (Department of Water Affairs & Forestry) (2008b). Consolidated water sector report: Fourth Quarter 2007/08. Unpublished report. Department of Water Affairs and Forestry, Pretoria

DWAF (Department of Water Affairs & Forestry) (2008c, July). Water quality management in the Free State. (Unpublished presentation to the Free State Water Quality Management Committee by Corinne de Kock, 31 July 2008)

DWA (Department of Water Affairs) (2010) Green drop report 2009. Version 1. South African waste water quality management performance. Department of Water Affairs, Pretoria

Eales K, Schreiner B (2008, August) Investing in infrastructure and institutional development for water security, growth and poverty reduction: some reflections on South Africa's learning. Paper presented at the World Water Week, Stockholm

EBRU (Environmental Biotechnology Research Unit) (n.d.). Integrated algal ponding systems: wastewater treatment, recovery and re-use in community social benefit applications. Unpublished memo produced by the Rhodes University Environmental Biotechnology Research Unit

ECDOH (Eastern Cape Department of Health) (2008, July). Report to the Eastern Cape provincial cabinet committee on the Ukhahlamba District Baby Deaths. (Paper presented). Retrieved December 17, 2008 from http//www.ecdoh.gov.za/uploads/files/170708161707.pdf

FinMark Trust (2006) Housing entrepreneurs: final report on small-scale landlords. Retrieved November 23, 2007 from http://www.finmark.org.za

Hausman R (2008) Final recommendations of the international panel on growth. Paper presented to National Treasury, South Africa Retrieved December 1, 2008 from http://www.treasury.gov.za/publications/other/growth/01-Overall percent20Summary percent20and percent20Final percent20Recommendations/01-Final percent20Recommendations percent20of percent20the percent20International percent20Panel.pdf

Hendricks L (2007, March 27). Speech presented at the handover of the Elundini local municipality bucket eradication programme. Retrieved August 15, 2007 from http://www.polity.org.za/article.php?a_id=106547

Hendricks L (2008a, May 23) Budget vote speech by Minister of Water Affairs and Forestry. Retrieved August 20, 2008 from http://www.dwaf.gov.za/Communications/MinisterSpeeches/2008/BudgetVote23May08.doc

Hendricks L (2008b, March 11) Water crisis in our country. (Speech by Minister Lindiwe Hendricks, Minister of Water Affairs and Forestry to Parliament. Retrieved August, 2008 from http://www.search.gov.za/.../previewDocument.jsp?... percent2Fdata percent2Fstatic percent-2Finfo percent2Fspeeches percent2F2008

Hetherington S (2008) Finding a home in the city. SADelivery, Edition 14, String Communications http://www.sadelivery.co.za/files/back_issues/delivery/Edition14/Housing_urban_townships0402.pdf

Ismawati Y (2007) SANIMAS: sanitation by the community in Denpasar, Indonesia. Report for the United Nations Economic and Social Commission for Asia and the Pacific & Pro-poor Water and Wastewater Management in Small Towns. Accessed 2 October 2008. www.unescap.org/pdd/prs/ProjectActivities/.../Water/.../Denpasar_ES.pdf

Lawless A (2007) Numbers and needs in local government. Civil engineering: the critical profession for service delivery. South African Institute of Civil Engineering

Lawless A (2008). Personal communication, 29 January 2008

Mafereka J (2007, September 30). Free state provinceresentation: Bucket eradication (Status). (Presentation by Mr Joel Mafereka, MEC for local government and housing, free state province' to the parliamentary portfolio committee on water affairs). Retrieved May 17, 2008 from http://www.pmg.org.za/minutes/20071009-bucket-eradication-programme-access-water-targets-progress-briefings

Molewa E (2008) North West Province Premier Edna Molewa, State of North West Province Address. 22 February 2008. Available at www.polity.org.za/article.php?a_id=127169

Momba MNB, Osode AN, Sibewu M (2006) The impact of inadequate wastewater treatment on the receiving water bodies – case study: Buffalo City and Nkokonbe Municipalities of the Eastern Cape Province. Water SA 32(5):687–692

Momba M, Obi CLI, Thompson P (2008) Improving disinfection efficiency in small drinking water treatment plants. Water Research Commission report, WRC 1531/1/08. Pretoria

Munnik V (2008) Creating wealth in rural and peri-urban communities through small wastewater treatment plants. Unpublished background paper, 31 July 2008, commissioned by the Mvula Trust

National Treasury (2008a) Conditional grants transferred from national departments and actual payments made by municipalities, Second quarter ended 31 December 2007. Retrieved November 20, 208 from http://www.treasury.gov.za/legislation/mfma/media_releases/section_71_2nd_0708/14. percent20FS percent20- percent202nd percent20Q percent20CG percent20- percent2015 percent20Feb percent202008.pdf

National Treasury (2008b, August 26) Local government budgets and expenditure review, 2008. Meeting new challenges of growth and poverty alleviation. Retrieved November 20, 2008 from http://www.treasury.gov.za/publications/igfr/2008/lg/Executive percent20Guide percent20-percent2026 percent20August percent202008.pdf

Ndlovu D (2007) Parliamentary submission on the eradication of the Bucket Toilet Syste, 23 & 23 February 2007. South African Local Government Association. Available at www.pmg.org.za/docs/2007/070223salga.ppt

NISC (2008) Communicable diseases communique, 2005–2008, monthly National Institute for Communicable Diseases. http://www.nicd.ac.za/

Office of the Presidency (2008) Towards a fifteen year review. South Africa. Retrieved December 09, 2008 from http://www.thepresidency.gov.za/main.asp?include=docs/15year/main.html

Omigeer K (2007) Briefing on the eradication of buckets by the Gauteng Provincial Department of Local Government to the Parliamentary Portfolio Committee on Water Affairs and Forestry, 4 May 2007. Available at www.pmg.org.za/minutes/20070503 eradication-buckets-water-and-sanitation-service-briefings-provincial-water-departm

Paterson C, Mara D, Curtis T (2007) Pro-poor sanitation technologies. Geoforum 38:901–907

Reiter P (2008, August 20) Practical implications and opportunities of climate change for water. Presentation made at World Water Week, Stockholm), 20 August 2008

RSA (1996) The Constitution of the republic of South Africa (Act No. 108 of 1996). Government Printer, Pretoria

RSA (2001) White paper on basic household sanitation policy. Government Printer, Pretoria

Savage D, Eales K, Smith L (2008) Securing South Africa's water sector to support growth and sustainable development. Unpublished paper. The Mvula Trust

Snyman HG (2007, June) Wastewater treatment in South Africa: the status quo. Unpublished presentation delivered at the WISA Wastewater Management Conference, 'From Challenge to Opportunity', East London

StatsSA (2005) Achieving a better life for all. Progress since Census '96 and Census 2001. Report no. 03-02-16, Statistics South Africa

StatsSA (2007) Community Survey 2007. Statistical Release, P0301, Statistics South Africa

Timm J (2008) Personal communication, 10 October 2008

Ukhahlamba District Municipality (2008, June) Water pollution and water challenges in Ukhahlamba District Municipalities. Presentation to the Parliamentary Portfolio Committee Public Hearings on Water Quality. Retrieved September 7, 2008 from http://www.pmg.org.za/files/docs/080603ukuhlamba.ppt

Ulrich A (2008) Personal communication, September–October 2008

UNDP (2008) Human development report 2007/2008. United Nations Development Programme. Retrieved from http://www.hdrstats.undp.org/indicators/5.html

UNESCO (2006) World water development report 2. United Nations Education, Scientific and Cultural Organisation, Paris.

Van Vuuren L (2006) Potential health time-bomb ticking in Free State. The Water Wheel, May/June 2006. ISSN 0258-2244

WHO (2008) Global tuberculosis control: surveillance, planning, financing. World Health Organisation. Retrieved December 17, 2008 from http://www.who.int/tb/publications/global_report/2008/en/index.html

Chapter 5
Transforming Legal Access to Water to Redress Social Inequity and Economic Inefficiency

Gavin Quibell, Robyn Stein, Ashwin Seetal, and Noxolo Ncapayi

Abstract The onset of democracy in South Africa provided the opportunity to make sweeping changes to legislation to allow for the proactive reallocation of water to address the effects of past racial discrimination. This is done through compulsory licensing provisions in the National Water Act. However, this process must balance the need to address social inequity, inefficiency in water use, economic growth and environmental sustainability with the rights of existing water users and the contribution they make to the economy and job creation.

Water reallocation reform can realise race and gender targets, and widespread reallocations to large numbers of the rural poor may make radical inroads into reducing rural poverty. Equally, redress is important for social and political stability. Key lessons have emerged from South Africa's approaches to balancing these considerations which are addressed in this chapter.

Keywords Water allocation • Water for economic growth • Beneficial use • Public interest • Water legislation • Race and gender equity • Water use authorisation • Water reallocation • Bill of Rights

---

G. Quibell (✉)
Independent Consultant, Pretoria, South Africa
e-mail: gavinquibell@yahoo.co.uk

R. Stein
University of the Witwatersrand, University of London, Edward Nathan Sonnenbergs Inc.
e-mail: rts@enf.co.za

A. Seetal
IWRM Specialist. Golder Associates Africa (Pty) Ltd, Pretoria, South Africa
e-mail: aseetal@golder.co.za

N. Ncapayi
Department of Water Affairs and Forestry, Private Bag X313, Pretoria 0001, South Africa
e-mail: ncapayin@dwaf.gov.za

B. Schreiner and R. Hassan (eds.), *Transforming Water Management in South Africa: Designing and Implementing a New Policy Framework*, Global Issues in Water Policy 2,
DOI 10.1007/978-90-481-9367-7_5,

5.1 Introduction

The onset of democracy in April 1994 provided the opportunity for South Africa to make sweeping changes to its legislation, not only to address the legacy of apartheid, but also to bring the country's water laws in line with some of the most up to date thinking on natural resources management. A long history of water resource management (see Chapters 1 and 2) and interactions with countries around the globe left South Africa with an experienced and well-qualified core of water resource managers. This, together with the new political dispensation, contributed to the formulation of a radically different water law, which not only allowed for proactive actions to redress the effects of past racial discrimination, but which also included cutting edge global thinking into the statute.

The result of this water law review process – initiated in 1994 by the then Minister of Water Affairs and Forestry, Prof Kader Asmal – was development of the National Water Act (Act 36 of 1998) (Republic of South Africa 1998). This Act now allows the radical transformation of legal access to the water to effect race and gender reform, but also to address economic inefficiency. As a result, the National Water Act (NWA) is widely regarded as one of the best of its kind in the world (Stein 2002).

The NWA does this primarily via the compulsory licensing provisions contained in Sections 43–48 (Republic of South Africa 1998). Compulsory licensing is a mechanism to reconsider all the water use authorisations in an area to achieve sweeping changes in the allocation of water. This process calls for all water users within a specified area to reapply for their water use entitlement.[1] This allows a fairer reallocation of the water, but can also promote more beneficial use of water, facilitate management of the resource, and can provide for protection of water quality and aquatic ecosystems (NWA – Section 43(1)). As a result, the compulsory licensing process is often seen as the panacea for all water resource management problems, and the end point of the Integrated Water Resource Management (IWRM) process.

However, in spite of the enormous opportunities provided by these provisions, compulsory licensing is yet to be implemented.[2] The challenges that allocation of scarce natural resources poses, particularly in the light of the need to balance social redress with sustained economic growth, as well as the deeply political nature of the process, seems to lie at the heart of this hold-up. Water allocation reform[3] (as with land reform) requires reallocation from the "haves" to the "have-nots." This not only has to balance the rights of the affected individuals with the benefits to the nation as well as those acquiring new water rights, but also holds the risk of impacting on the economy, especially where new enterprises may struggle to establish. Water allocation reform must also recognise

[1] In South Africa "water rights" are referred to as "water use entitlements," to distinguish the basic right to water contained in the Reserve (see Chapter 5) from the right to use water conferred by the NWA.

[2] The Department of water affairs has recently (2010) promulgated compulsory licensing in three areas.

[3] The term water allocation reform is used in this chapter to distinguish it with water reform, which could include improved access to potable water and sanitation (see Chapter 6). Water allocation reform refers to transforming legal access to water for productive or commercial purposes.

that only a small proportion of South Africans will be using water for productive or commercial purposes, the overwhelming majority would derive benefits from economic growth that may be linked to water allocations to other enterprises. In addition, water allocation reform, perhaps more than any of the other programmes to implement the NWA, holds a high potential for legal action and must be grounded in the Constitutional provisions for redress and must be a consistently fair and just process.

South Africa's Department of Water Affairs (DWA) has therefore spent considerable effort in developing approaches for compulsory licensing that ensure a carefully considered process which takes positive steps towards race and gender reform, but which minimises the risks to the economy. Much has been learnt on how these provisions can be used to realise race and gender reform, address poverty, promote economic growth, and minimise the potential for legal action.

In this chapter we explore the legislative provisions for reallocating water to address the legacy of apartheid, as well as the opportunity these provisions provide for maximising the social and economic returns from water use. However, we also explore the challenges and risks this process holds, and how these can be addressed in the compulsory licensing process. This analysis might resonate for those countries facing absolute water stress and basin closure, where the reallocation of water is the only way to support continued growth while maintaining social equity. This may become even more poignant as climate change places further stresses on water availability.

5.2 Legislative Provisions for Redress

5.2.1 The Constitution

The Constitution of the Republic of South Africa (Act 108 of 1996) is the basis for the country's democracy and, as the supreme law of the country, also provides the basis for transforming legal access to water. The first clues as to how government should tackle redress come from the Preamble to the Constitution, which reads as follows;

> We the people of South Africa,
> Recognise the injustices of our past;
> Honour those who suffered for justice and freedom in our land;
> Respect those who have worked to build and develop our country….
> (Republic of South Africa 1996)

At the outset, the Constitution, therefore, acknowledges that we must address the injustices of the past, but also recognises the economic development of the country, and affirms the rights of <u>all</u> South Africans to a just reform process. This is given substance in Section 9(2) of the Constitution, which requires government to develop legislative and other measures to promote equality, and is given specific emphasis in the property clause of the Bill of Rights, which allows for the expropriation of property in the public interest [S25(2*a*]. The public interest, in this context, includes reforms to bring about equitable access to South Africa's natural resources [Section 25(4)*a*]. Section 25(8) further indicates that no provision of the property clause may impede the State from undertaking land, *water* and related reform.

The Constitution consequently explicitly recognises the need for reallocating water to affect social inequity in the property clauses of the Bill of Rights. Van der Schyff (2004) argues that, in the post Constitutional dispensation, water use must be regarded as a "rights in property" and is consequently subject to the provisions of Section 25 of the Constitution dealing with compensation. This argument is built on in the National Water Policy, which recognises that the reallocations of water will be protected by the Constitutional provisions for corrective action (DWAF 1997). The NWA therefore recognises that where the user suffers severe economic prejudice as a consequence of water entitlements being curtailed, that compensation is to be calculated according to Section 25(3) of the Constitution[4] (NWA, Section 22 (7)*a*).

However, recognising that the Bill of Rights also allows for the limitation of these rights (Section 36), the NWA also indicates that the calculation of compensation must be determined by disregarding the water taken away to rectify an unfair or disproportionate water use, to rectify an overallocation, or to provide for the Reserve.[5] These provisions, however, do not allow government to arbitrarily expropriate water, and Section 25(2) indicates that property (which must include water) may only be expropriated in the public interest. The public interest includes actions to bring about reform.

The highest law in the land therefore requires government to introduce legislative provisions for water allocation reform. These provisions should promote redress, social and economic development, but without compromising environmental sustainability. However, government is also required to balance the benefits to the nation with the rights of the individual. Steps to transform legal access to water may not be arbitrary and are only justified when the reallocation of water is in the best interests of the nation as a whole. Together these provisions provide the basis for the compulsory licensing provisions of the NWA, but also indicate how these provisions must be implemented.

5.2.2 Water Use Entitlements in the National Water Act

The purpose of the NWA is to ensure that the nation's water resources are protected, used, developed, conserved, managed and controlled in order to *inter alia*:

- Provide for water needs now and into the future.
- Promote redress and equitable access.
- Promote efficient, sustainable and beneficial use of water.
- Protect aquatic and associated ecosystems.
- Reduce and prevent pollution.

---

[4] These provisions of the NWA have attracted considerable criticism, and have yet to be tested in the courts. The approach currently advocated in the DWAF is however that the provisions of the NWA must be implemented as they stand.

[5] The Reserve is the quantity and quality of water required for basic human needs, as well as for ecological functioning (see Chapter 5).

[Republic of South Africa 1998 – **Section 2** – paraphrased here]

The need for equitable access to water resources, as well as the efficient and beneficial use of water therefore lies at the core of the NWA. The key to securing these goals lies in transforming legal access to water use via water entitlements or water rights as outlined in Section 22(1).

These entitlements include;

- **Schedule 1**[6] use – small volumes of water or wastewater used at a household level with little potential for negative impacts on the water resource, for which no application or authorisation needs to be made.
- **General Authorisations** – larger volumes of water or waste with some potential for negative impacts on the water resource, which may be generally authorised for a specific water resource, specific group of users, or type of water use.
- **Existing Lawful Use** – which is a water use that lawfully took place in the period two years before the commencement of the NWA, and
- **Licensed Water Use** – larger volumes of water, waste or other water use authorised in terms of a licence issued under the NWA, and upon approval of an application.

Existing lawful water use is a temporary bridging measure, to be used until the entitlement can replaced by a licence under the compulsory licensing process. More importantly, existing lawful water use, because it was established under discriminatory legislation does not allow for race and gender equity, neither does it necessarily accommodate the needs of the environment.[7] Compulsory licensing, therefore, aims to shift the predominantly white and male dominated existing lawful water use – into licences and general authorisations that make provision for race and gender shifts in water use, but also to promote efficiency and ensure environmental sustainability.

Importantly, the provisions of Section 27 of the NWA must be taken into account when considering any general authorisation or licence to use water even under the compulsory licensing process. This requires the licensing authority to *inter alia*;

- Accommodate the requirements of the Reserve and Class (see Chapter 6);
- Redress past racial and gender discrimination;
- Support the efficient and beneficial use of water in the public interest;
- Consider any catchment management strategy applicable;
- Consider the impacts on existing lawful users of water; and
- Consider the investments made by existing users as well as the investments to be made to use the water.

---

[6]All South Africans have access to Schedule 1 use. This allows for the small-scale use of water, not for commercial purposes, to improve livelihoods (this may include food security as well as the trading of produce for small gains to support basic needs). The focus of transforming legal access to water via compulsory licensing is for uses over and above Schedule 1 use.

[7]Many argue that the old Water Act of 1956 was not discriminatory per se, however, as access to water was tied to land, this Act was in effect discriminatory. Moreover many allocation decisions made under the old Water Act did make provision for environmental flows.

5.2.3 Compulsory Licensing

The provisions for compulsory licensing in the NWA (Sections 43–40), allow for the proactive reallocation of water. Compulsory licensing may be instituted to;

- Achieve a fair allocation of water from a resource that is under stress, or to achieve equity in allocations;
- Promote beneficial use of water in the public interest;
- Facilitate efficient management of the water resource or;
- Protect water resource quality.

While any of the 11 uses of water contemplated in Section 21 of the NWA could be subject to compulsory licensing, initially only the abstraction of water, water storage and stream flow reduction activities[8] were to be targeted for compulsory licensing (DWAF 2006a).

The compulsory licensing process (Fig. 5.1) is initiated by publishing a notice in the Government Gazette calling for persons who wish to use water for commercial purposes (i.e. outside Schedule 1) to apply for licences. This notice will require all existing and new water users within any given geographical area, to apply at a

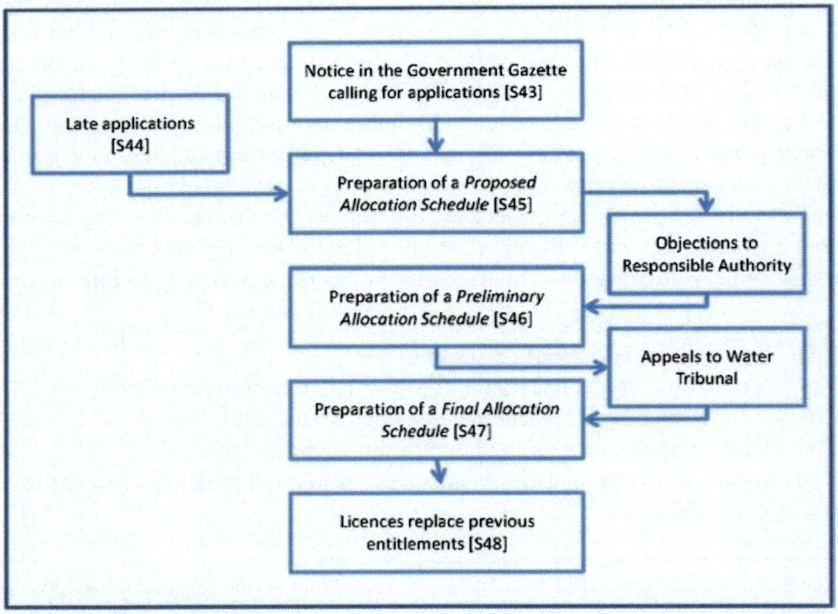

Fig. 5.1 The Compulsory Licensing process outlined in the National Water Act

[8] **Section 36 1b** of the NWA allows the Minister to declare certain activities as Stream Flow Reduction Activities (SFRAs). These are activities that reduce runoff, but do not entail abstraction of surface or ground- water. To date only commercial afforestation has been declared as an SFRA [**S36 1 a**].

specific address and within a specified timeframe [S43] (late applications can be considered [S44]). Such an application may attract a licence application fee, or late application fee.

Once all the applications have been received a *proposed allocation schedule* must be prepared, which specifies how the water is to be allocated to each of the applicants. It is not necessary to allocate all the available water [S45(3)], and some may set some aside for future use. This proposed allocation schedule must reflect the quantity of water to be;

(a) Assigned to meet the requirements of the Reserve and any relevant international requirements;
(b) Assigned to meet the requirements of existing licences;
(c) Allocated to applicants to redress the effects of racial and gender discrimination;
(d) Allocated to applicants exercising existing lawful water uses, to whom the licensing authority determines licences should be issued;
(e) Allocated to each of the applicants, taking into account the requirements of Section 27; and
(f) Allocated to every other applicant by public tender or auction

NWA, Section 45(2)

The outcome of this reconciliation process is a proposed allocation schedule. This must be published for comment [S45(4)], and stakeholders may raise objections to the proposals. Objections received must be considered, and if found to hold merit, must be accommodated into a preliminary allocation schedule [S46(1)]. Stakeholders who raised objections to the proposed water allocation schedule, may then (if still not satisfied) raise appeals to the preliminary allocation schedule to the Water Tribunal. After hearing appeals a Water Tribunal may direct the amendment of the preliminary allocation schedule [S46(2)]. A preliminary allocation schedule becomes final if no appeals are lodged, if the schedule has been amended following successful appeals, or if the appeals are dismissed [S47]. Once the allocation schedule becomes final, licences must be issued according to the schedule [S47(2)], and these licences replace the previous existing lawful use entitlements [S48].

Box 5.1 Curtailing Water Users Already Licensed Under the NWA

Licensed water use can be curtailed under **Section 49**, which allows a responsible authority to undertake a general review of licences, and to amend any condition of these licences. The review of licences can not shorten the period of the licence [S49(2)], and the general review of licences can not remove an entitlement in its entirety. Moreover, as the requirements of Section 27 have already been taken into account licence holders may have stronger claims for compensation if their water use is curtailed. These provisions also allow for an ongoing review and reallocation process as conditions change. Some water users have indicated that this affects their willingness to invest in new infrastructure.

The reconciliation process therefore moves towards the allocation of water in a way that realises the need for social equity and economic growth, but also in a way that recognises the interests of the affected water users.

5.2.4 General Authorisations in Compulsory Licensing

The compulsory licensing provisions outlined in the previous sections are not inherently pro-poor. They require people to apply for water, within a specific period, and at a venue that may be far from where they live. Moreover, the application should attract an application fee (to avoid opportunistic applications). This can place a considerable burden on the poor, who not only have to make the application in a non-mother tongue language, but must also have a clear idea of how much water they require, and for what purpose.

However, the provisions of Section 45(3) can be used to set water aside water for potential uptake by the rural poor. Water set aside in this manner can be generally authorised under Section 39 of the NWA. This allows the general authorisation of any category of persons to take up the water without the need for a licence.[9] A general authorisation is established by publishing a notice in the Government Gazette, and inviting comments from interested parties [S39(4)(a)]. This general authorisation does not limit any other entitlement to use water [S39(5)] and as such people covered by the authorisation can still apply for a licence if they wish. General authorisations are also subject to the requirements of Section 27, and are established for a specified timeframe, where after they may be extended or withdrawn.

General authorisations can therefore ring fence water set aside during compulsory licensing for race and gender reform, and limit the burden of application for the rural poor. This can also reduce administrative burdens, and can limit opportunistic applications. More importantly, this allows for the gradual uptake of the water by the rural poor, supported by appropriate empowerment programmes in the post compulsory licensing phase.

5.3 Approaches to Reallocation of Water

5.3.1 Beneficial use in the Public Interest

The underlying principles of the Constitutional requirements for reform, the public trust principle outlined in the National Water Policy, and the provisions of the NWA are that reform initiatives must be in the best interests of the whole nation. South Africa's National Water Policy indicates that water use will not be recognised unless it reflects this 'beneficial use in the public interest,' and that beneficial use

[9] Water users taking up a general authorisation may still be required to register their water use

in the public interest means the optimum balance between social, economic and environmental use of water (DWAF 1997).

The NWA does not define beneficial use in the public interest. However, the Department of Water Affairs and Forestry's "Framework for Water Allocation" (DWAF 2006a) outlines beneficial use as;

> ...water allocations that are to the benefit of the public and the nation as a whole. It balances the broader pubic interest with the rights of the individual, and includes the commitment to equity

This reflects the Constitutional requirements for equity, but also the need to recognise the rights of the individual water users and their contribution to the economy. DWAF's Toolkit for Water Allocation Reform (available at http://www.dwaf.gov.za/WAR/doctoolkit.asp) further elaborates beneficial use of water as including considerations for economic growth and social development, job creation, equitable access to water for productive purposes, social stability, investor confidence, protection of aquatic ecosystems, efficient and non-wasteful water use, and a balance between the water using sectors. (DWAF 2006b).

This broader concept of beneficial use recognises that not everyone will get water for commercial or productive purposes, but rather that the economic benefits of the water use should be more equitably spread. The Water Research Commission (1996 and 1999) estimate that there are some 200,000 to 250,000 small-scale irrigation farmers in South Africa, of which the majority are black women. If this number of users is doubled through the water allocation reform process, and if each of these users live in a household of six (average size of a rural household), this small-holder irrigation water use will directly impact on the lives of some 3 million South Africans. Assuming a threefold knock on benefit to this use, only 20% of an estimated 45 million people in South Africa would directly benefit from commercial irrigation. The remaining 80% would have to derive secondary benefits from a growing economy through job creation.

However, beneficial use in the public interest also inherently recognises that the economic benefits of water use in terms of jobs and income per drop must be balanced with political and social objectives for redress and hence social stability. Moreover, it recognises the need for a diverse and stable economy, and investor confidence. This means that it is not only how much water changes hands, but also that the way in which this is done is important. It is, nevertheless, also important that positive steps to achieve equity can promote social and political stability. It is therefore not sufficient to simply leave water in white hands on the basis that benefits will best accrue to black people and women through economic use by white-owned enterprise, and direct uptake of water by black enterprises is essential.[10]

Compulsory licensing may, therefore, result in a number of difficult reconciliation choices. The following sections outline a dichotomy in approaches for realising

---

[10] In spite of this, of the some 1,000 licences issued between 1998 and 2007, very few were issued to black users or communities. While some applications from black owned enterprises are still awaiting processing, the majority of requests for licences are still received from white and male applicants (most often for water trading).

beneficial use in the public interest, highlighting the risks and advantages of these by drawing on examples from South Africa.

5.3.2 Distributive or Growth Models for Reallocations?

Two dichotomous ideologies can drive the reallocation of water to realise social redress and economic efficiency. The first emphasises fewer large allocations to larger black or women owned enterprises to promote economic growth, incomes and jobs per drop, "*the redistribution for growth model*," and the second emphasises the reallocation of smaller amounts of water to large numbers of the rural poor (possibly as general authorisations), "*the distributive model*."

Importantly, the *growth model* does not mean that water remains in white hands, but rather that the emphasis would be on shifting some water into larger black and women owned enterprises offering economically efficient water use and job creation. This requires that these enterprises make application during the compulsory licensing process. Similarly, the "*distributive model*" does not imply less efficient water use and fewer jobs per drop, as smaller family owned farming units are often more productive (Van Zyl et al. 1995). Moreover, the distributive model does not rely on the applications for licences as smaller amounts of water could be set aside and generally authorised for uptake by black and/or women users.

The *growth model* is characterised by fewer larger water users, who are employers (not necessarily in labour intensive enterprises), and may be supported from large storage schemes providing a higher assurance of supply. Larger scale water users also tend to buy in support or other services (for example technical or commercial support), helping maintain a more diverse economy. Conversely, the *distributive model* is characterised by large numbers of smaller water users, perhaps in family owned and run businesses.[11] These users may be dependent on smaller schemes or run of river and groundwater abstractions, and may not have significant storage capacity. This, together with more marginal profit margins may make these users more vulnerable, and hence potentially to the impacts of climate variability, as they may not be able to survive periods of water shortages.[12] Smaller water-using enterprises may also not buy in support services and their contribution to the economy may be limited to the direct sale of their produce. In these cases local economic multipliers may be limited, and the benefits of water use may not be widespread.

The equity aspects of the *growth model* is predicated on the assumption the benefits of water use rather than water use per se are spread, either through employment and/or through local and national economic multipliers. Larger enterprises also provide the opportunity for promoting equity ownership schemes for

---

[11] The discussion here focuses on the small-scale commercial use of water, beyond that which is secured as Schedule 1 use, and therefore that water use which will be subject to compulsory licensing.

[12] Kabat and van Schaik (2003) show that climate change may result in between 5% and 50% decreases in water availability for much of South Africa.

employees. Equity partnerships are becoming more common in agricultural support programmes in South Africa, and experience has shown that equity shareholding arrangements (stimulated by access to water) often give better returns to beneficiaries than primary use of the water resource. The reasons for this are diverse but typically primary water use is often higher risk and is more dependent on a range of uncertain external factors including access to finance, markets and production information with a greater need for management skills. Importantly, sugar and forestry are two exceptions with respect to this smallholder risk (Mayson 2003).

However, equity ownership schemes do not always benefit the poorest, and attempts to promote redress by equity ownership in the Blyde River irrigation scheme in the Limpopo Province (see Box 5.1), lead to existing white owners selling shares to larger black owned enterprises. In this case benefits of water use would not necessarily have spread to poor communities in the area.[13]

The *growth model*'s contribution to spreading the benefits of water use, and hence to widespread poverty eradication is also to some extent dependent on broader economic growth in the country. Poonyth et al. (2001) indicate that industrial growth in South Africa requires an expanding agricultural sector, suggesting that the benefits of water use can be felt far beyond the primary water use sector. Active promotion of the agricultural sector may therefore stimulate broader economic growth and hence addressing poverty. In this way the benefits of water use, even within a few larger enterprises, may be realised by more people. Economic growth will, however, be influenced by a number of other factors, not least of which could be global energy costs. Geo-political tensions may consequently slow economic growth, limiting this spread of the benefits of water use.

The South African government is aiming at an average economic growth rate of 4.5% until 2009, and 6% thereafter (South Africa Info 2006), recognising that this would start to generate sufficient employment to make sustainable inroads into job creation and poverty. However, while the country's economy has recently fared remarkably well, these growth targets may be threatened by global events and recessions. Importantly, in spite of this positive economic growth since 1994, formal employment in South Africa has continued to decline (www.wikipedia.org/wiki/Economy_of_South_Africa). In the face of slower economic growth, the reallocation of water to larger numbers of primary water users (the *distributive model*) may therefore ultimately benefit more people.

Moreover, the benefits of economic growth, while perhaps more sustainable and less risky, tend to filter slowly down to the rural poor, leading to frustration at the slow pace of reform. Promoting uptake of water by larger black owned enterprises may, at least initially, further increase the gap between the "haves" and the "have-nots."[14] In these cases the benefits of water allocation reform will not necessarily spread to the poor, or contribute to redress, particularly if national economic growth

[13] The Blyde 800 scheme is presently "on ice," pending the outcome of land claims by people disposed of land under the apartheid system.

[14] Statistics in South Africa show that the gap between rich and poor is widening in spite of the successful promotion of Black Economic Empowerment. This lead to the promotion of the concept of Broad Based Black Economic Empowerment – which aims at a wider set of beneficiaries.

remains slow. For this reason, macroeconomic policies supporting strong economic growth, rather than redistribution, do not always attract wide support from grass-roots movements. More importantly, when these policies fail to deliver on the promise of a better life for all, social pressures for more radical redistribution policies and service delivery grow.

This radicalisation of approaches for redress has already occurred within South Africa's land reform programme, where the slow pace of delivery has promoted a rethink of the willing buyer – willing seller approach to land reform. The Restitution of Land Rights Amendment Act of 2003 (Republic of South Africa 2003) now gives more powers to the Minister of Land Affairs to expropriate land, and these powers are increasingly being exercised.[15] The Centre for Development and Enterprise (CDE 2005) has also suggested that popularist land rights movements will gain increasing support in the face of a slow pace of land reform. Similar shifts towards a more *distributive model* for water allocations may be expected if a predominantly *growth model* for water allocation reform programme fails to deliver benefits in the short-term.

Some proponents of the *growth model* argue that larger enterprises can make use of economies of scale represent a more economic use of water. Evidence, nevertheless, suggests that the *distributive model* may not imply lower productivity and economic growth, and a number of authors have indicated that smaller enterprises may be more productive than larger units (in van Zyl et al. 1995). Similarly, water use efficiency may be higher in smaller enterprises. The *distributive model* can also be effected via general authorisations, freeing up individuals and the responsible authority from the administrative burdens of licensing. General authorisations can also mirror customary laws, making implementation more effective in the rural areas, and allow potential black and women entrepreneurs to emerge slowly, taking their first step in becoming larger commercial users in their own right. Importantly, it does not preclude these individuals from applying for licences for larger volumes of water at any time.

The success of small scale irrigation enterprises established through general authorisations is, however, often dependent on other factors. The dismantling of the marketing board mechanism for agricultural products following the recommendations of the Kassier Commission (Kassier 1992), and a new government trying to appease the World Trade Organisation, has left South Africa with one of the most deregulated agricultural sectors in the world. The effects of this on agriculture have been dramatic, over the last 6 years the number of dairy farmers has more than halved, while those who have stayed in business have more than doubled production. Some 40% of the country's tomatoes come from one producer, and similar effects have been noted with other products. Clearly, establishing and maintaining small agri-business in South Africa, in the face of dwindling government support is becoming increasingly risky (CDE 2005 and Box 5.2).

Another key benefit to the *distributive model* is, nevertheless, that larger numbers of the rural poor get water entitlements during compulsory licensing (either as licences or general authorisations) and can start using this water to improve livelihoods beyond

[15] Interestingly, Land Reform also shifts the water that was associated with that land into black hands. This may in fact prove a more effective means of securing water allocation reform.

that water available as Schedule 1. Moreover, the majority of small-scale irrigation farmers are women. This gets the redress numbers right in the short-term, and addresses the immediate social and political pressures for reallocating water. DWAF 2006b identifies a number of factors that contribute to productive water use;

- The mandate to the land i.e. ownership of or permission to use or occupy the land;
- Financial resources – i.e. the funds for infrastructure and operation and maintenance;
- Technical skills and extension support;
- Markets for the products of the water use;
- Appropriate institutional arrangements;
- Planning skills – i.e. the ability to plan for the water use, and manage shortages;
- Enthusiasm – i.e. the desire to use the water;
- Security for the water using infrastructure and products;[16]
- Sense of catchment – i.e. the recognition that the use forms part of a wider catchment, and is affected by upstream use and effects downstream use and the aquatic ecology.

Water allocations and compulsory licensing can only address a few of these aspects, and while government programmes and financial support mechanisms are available, implementing these mechanisms relies heavily on cooperative governance. Inter-governmental cooperation is still notoriously difficult to achieve in spite of high-level government intervention and the Intergovernmental Framework Relations Act (Act 13 of 2005) (Republic of South Africa 2005). The Department of Provincial and Local Government's 10-year review has indicated that cooperative governance is largely informal, and has very high transactional costs (DPLG 2004). There are consequently numerous examples of schemes failing due to a lack of effective coordination and support (See Box 5.2).

However, there are examples of successful small-scale enterprises where private enterprise has become involved;

- The sugar industry has established 48,000 small (mostly black) users over the last 30 years;
- A further 78,000 ha of land under sugar will be transferred by 2014 bringing the total redistribution of land under sugar to 30%;
- The timber industry has established outgrower arrangements with some 10,000 emerging producers;
- One major cotton producer works with 1,500 small growers on traditional land; and
- The South African Breweries have promised 180 emerging farmers in the Taung area that they will buy their crop (CDE 2005)

The key element of successful establishment of these small-scale irrigation businesses has been that there was a ready market for the product. Markets for products have been identified as being critical to the establishment of viable water using

[16] There is increasing evidence that crime is affecting the sustainability of many land and water reform projects.

Box 5.2 Small Scale Irrigation Challenges

Revitalisation of irrigation schemes in Limpopo province (RESIS). Attempts to revitalise defunct irrigation schemes in the province are failing due to excessively bureaucratic approaches, and lack of attention to the established principles for successful irrigation schemes. (De Lange 2006)

Irrigation schemes in the Free State province. Water has been made available to support emerging irrigation farmers in the Free State province, but the virtual collapse of the Provincial Department of Agriculture means that this water has not been taken up. (van der Merwe 2006)

The Mhlathuze catchment. 2,000 ha of irrigation is available for uptake by emerging farmers, but only 1100ha has been taken up because of a range of factors including poor access to markets, political infighting and lack of extension and financial support. (DWAF 2004)

The Hluhluwe catchment. Some 500 ha of irrigation was made available to emerging farmers during the construction of the Hluhluwe Dam but lack of attention to all of the elements required for productive water use has made it difficult for emerging users to take up this water. (Perkins 2006)

The Lower Orange Scheme. Large volumes of water have been made available for emerging farmers, but attempts to establish this use have been held up by lack of funding to establish the irrigation infrastructure and poor cooperation between government departments.

enterprises (Chancellor et al. 2003). However, redress via allocations to large numbers of smaller enterprises relies on entrepreneurial skills being present in large numbers of the poor, and more importantly that significant numbers of the poor would want to take up water beyond their Schedule 1 allocations. A study by the Centre for Development and Enterprise suggests that only 6% of black people not currently in farming wanted to farm, and only 15% of farm workers wanted to farm on their own (CDE 2005). This may suggest the uptake of general authorisations tailored for the rural poor may not be widespread. However, if only 6% of the rural poor take up water during compulsory licensing, the numbers of irrigation water users are still likely to more than double in most areas of the country.

The above discussions have largely centred on promoting the use of water in agricultural enterprises. However, this is not the only means of promoting uptake of water by previously disadvantaged South Africans, and the water use in black and women owned enterprises in other sectors may offer better prospects for redress. However, large-scale mining and industrial activities require significant start up costs and, while compulsory licensing should certainly support the water demands for these activities, they perhaps offer fewer opportunities for reallocating water directly to black users. However, commercial use of potable water within local government may provide good opportunities via Small, Micro and Medium Enterprises (SMMEs). Uptake of this water will be promoted primarily via Integrated Development Plans (IDPs), which outline local governments plans to promote local economic growth and service provision, and by the demand projections from water services providers (see Chapter 3).

5.3.3 Economies of Scale

One of the key aspects of the *growth model* is the potential for economies of scale to support higher and more sustainable economic growth in larger farming units. International experience has, however, shown that smaller farming units tend to be more efficient largely because of the advantages offered by family labour (Van Zyl et al. 1995). This seems to hold true for the African experience and a World Bank study (in Van Zyl et al. 1995) has shown that employment is 30 times higher on small farms (<0.5 ha), and outputs 19 times higher, than on larger units (>8 ha). While it may be argued that this does not necessarily hold for irrigation farming, black owned irrigated sugar farms in Mpumalanga produce an average of 116.8 t/ha on average farm size of 7.1 ha, as opposed to 102.9 t/ha on an average of 68.6 ha for their white counterparts (Van Zyl et al. 1995).

Van Zyl et al. (1995), indicate that the effects of economy of scale on farming enterprises are relevant with respect to;

- "Lumpy" inputs, for example harvesting machinery and irrigation equipment
- Easier access to credit facilities for larger farms, and
- Costs of processing the product and easier access to markets

Smaller irrigation enterprises are at a higher risk due to limited access to credit, the high costs of establishing irrigation infrastructure and access to markets. However, contract farming, which provides ready markets, and often technical and financial support avoids the problems of economies of scale, and perhaps explains the success of sugar and forestry outgrower schemes. The problem of access to finance can also be addressed via targeted subsidy schemes (see Chapter 9).

It is, nevertheless, important to distinguish between the economic efficiency and size of an irrigation enterprise, and the water use per unit area. While the above arguments suggest that smaller irrigation units may be economically more efficient, they will not necessarily use less water per unit area. Larger irrigation farms have the opportunity to expand the area under irrigation, with little or no increase in water use. Irrigation farmers with larger farms can consequently accept reduced yields over a larger area but still increase the total income to the enterprise. Curtailments to larger users in favour of smaller users via the *distributive approach* can consequently stimulate greater productivity, and the more efficient use of water.

The number of users that can benefit from the *distributive approach* will however be limited by the availability of water and only a small proportion of the rural poor are likely to benefit directly from this approach. The benefits of water use may also, consequently, only filter slowly down to the majority.

5.3.4 Blue and Green Water Allocations

Falkenmark (1995) introduced the concept of blue and green water. Green water is that water that is lost via evaporation or evapo-transpiration, and includes the water use by rain fed crops, commercial forests and natural vegetation. Blue water is that portion of the total rainfall on the catchment that contributes to surface water run-off, or to the recharge of groundwater. Typically blue water only accounts for a small potion of the total rainfall on the catchment (usually 6–10%), and allocation policies based only on this portion give a skewed picture of the total water available to promote redress and economic growth.

Promoting "green water use" (rather than blue water allocations) by encouraging rain fed crops, supported by rainfall harvesting and soil water conservation practices, can therefore provide a more complete picture of water use. This holds several advantages; it reduces or eliminates the infrastructure costs associated with irrigation, hence addressing the lumpy inputs of economies of scale, and allows for water use far from surface water resources. Green water use is consequently particularly suited to promoting the *distributive model* of water allocation for the rural poor. However, this approach will be limited to areas with sufficient rainfall, and will be particularly vulnerable to climate variability. In addition, monitoring of allocations based on green water use will require a re-think of the current approaches to water allocation.

While the NWA can accommodate green water use as SFRAs [NWA, S21*d*, and S36], only green water use by commercial forestry is currently considered as part of the water allocation and compulsory licensing processes. Although more activities could be declared as SFRAs, the practicalities of expanding the concept to all rain fed crops, and the potential regulatory impacts on the rural poor, are likely to limit the use of these provisions for redress purposes. This, however, does not preclude the responsible authority from accounting for and monitoring both blue and green water use when making, and reporting on, allocation decisions.

In this respect, the DWAF is actively promoting the use of rainwater harvesting; particularly to address basic food needs, by providing targeted subsidies for rainwater tanks. This water use is expected to have negligible impacts on blue water flows, and is authorised as Schedule 1 use. However, the expansion of the concept to small commercial 1–2 ha plots could further promote the use of green water "allocations" to address social inequity and the more efficient use of the total water available. Falkenmark and Rockström (2004) in fact argue that green water use by grasslands and forests must be converted into crop production in order to address growing global food needs.

In the drier regions of South Africa, this is likely to require soil water retention systems, or mechanisms to increase water retention. However, Calder (2005) reports on the Karnakta and Andra Pradesh Rural Livelihoods Projects in India, soil water retention structures to promote small scale irrigation from groundwater have increased total green water losses, reducing blue water flows for downstream users. There is, nevertheless, merit in broadening the water allocation picture to include both green water and blue water "allocations," and in further exploring the implications this may have for increasing social equity and the efficient use of water. Certainly, a more complete green water based picture of water use may influence decisions on the balance between the *growth* and *distributive models* of water allocation.

5.3.5 Summary

The call for licences under compulsory licensing is unlikely to realise applications that inherently reflect beneficial use in the public interest, and difficult reconciliation decisions will have to be made. Moreover, existing white users will have to be curtailed for equity purposes. Water allocations, and in particular decisions on how much water should be set aside as general authorisations, must therefore balance the *distributive or growth* ideologies for water allocations.

The Centre for Development and Enterprise report a similar tension within the land reform programme, where certain officials in the Department of Land Affairs favour the redistribution of land to large numbers of the rural poor, whereas others favour the allocation of larger parcels of land to fewer large commercial BEE enterprises (CDE 2005). Chapter 3 and 4 of this book also propose a similar tension in meeting water and sanitation targets, where the need to get the numbers right in the short term, must be balanced with the need to ensure sustainable water and sanitation provision. These tensions are likely to be inherent in many redress initiatives, including the global redress mechanisms outlined in the Millennium Development Goals. However, water allocation reform offers opportunities for balancing these approaches, perhaps not as evident in other reform initiatives.

5.4 Balancing the Approaches

5.4.1 Background

In the previous section we introduced two ideologies for reconciling demands for, and availability of, water during compulsory licensing. While these are presented as two separate approaches for realising the objectives of redress and economically efficient use, they are not mutually exclusive and an appropriate balance will have to be found in any catchment. This will not only be determined by the need for redress and economic efficiency, but also by both the type and number of applications received under the compulsory licensing process and the need to balance allocation decisions with national development objectives. In the following sections we show how the compulsory licensing process can be used to help find this balance.

5.4.2 Planning for Small-Scale Uptake

The Department of Water Affairs and Forestry's "Toolkit for Water Allocation Reform" (DWAF 2006b) outlines the processes that should support compulsory licensing, and proposes the development of a Water Allocation Plan[17] before the call for licences. This draft Water Allocation Plan would;

- Identify good opportunities for water use in the area, highlighting where markets for products are available, and where other assistance is available to support the uptake of the water.
- Highlight the constraints to uptake of some of these opportunities.
- Outline the water that may be set aside and/or generally authorised to support uptake in small enterprises.
- Propose empowerment programmes to support the application process, or the uptake of water in the general authorisations.
- Suggest possible curtailments to existing lawful use, and the point at which severe economic prejudice to these users may occur.
- Indicate the criteria that would be used for evaluating applications.
- Suggest possible water trading options between larger commercial enterprises.
- Outline options to parallel gradual reductions in existing use with the uptake of the water by emerging users under the general authorisations.
- Identify the benefits to regional stability and growth offered by supporting particular allocations.
- Highlight particular cautions for curtailing some users to avoid knock on effects on the regional economy.
- Outline the timeframes for the remainder of the compulsory licensing process, i.e. when the call for applications will be made, the closing date for applications, and projections for when the proposed, preliminary and final allocation schedules may be available.
- Indicate how and where potential applicants could submit applications.

The (draft) Water Allocation Plan would be developed in consultation with stakeholders, would provide a framework for the reconciliation process, and may help guide applications under compulsory licensing. This would not only influence the type of applications received, but will help identify the most appropriate water allocation scenarios. This Plan would consequently highlight where the best opportunities lie for uptake of water as either general authorisations and/or licences, and would provide a basis for balancing the *growth* and *distributive* approaches.

General authorisations would, therefore, only be established where the greatest opportunities and lowest risks for uptake of small volumes of water lie. General authorisations would typically support development initiatives by other development agencies and private enterprise, and would be more likely to attract the support needed to establish sustainable small water using enterprises. This would also provide those with the entrepreneurial initiative the opportunity to start up small-scale water use without the burden of having to apply for a licence within a specified timeframe.

---

[17] A Water Allocation Plan forms part of the Catchment Management Strategy, which is developed by a Catchment Management Agency, **[S9e]** of the NWA. If this plan has already been developed, it can form the basis for reconciliation in compulsory licensing. If not, a draft Allocation Plan guides the process.

Moreover, the risks of curtailments to the regional economy and job provision will be highlighted, and hence helping identify opportunities for equity sharing.

More importantly, paralleling the uptake of water set aside as general authorisations with gradual reductions by the existing lawful water users (and improved water use efficiency) can minimise the disruptions to the economy. In this way the risks of setting too much water aside for the *distributive approach* can offset against opportunities for existing users to gradually curtail their use.

5.4.3 Minimising the Impacts on Economic Growth

Uptake of water for social equity within the *growth model* is largely dependent on the number and nature of licence applications received. This, however, does not mean that all the applicants will get water, or even that all the black or women applicants will get water, and broader concept beneficial uses of water in public interest must still underlie allocation decisions. In this respect the Framework for Water Allocation (DWAF 2006a) recognises that curtailment of existing water use, particularly to the point at which the user may suffer severe economic prejudice, holds risks for the economy and increases the possibility of legal action. This could slow the reform process. The Framework therefore outlines a hierarchy of processes to consider before curtailing existing lawful water users to the point at which severe economic prejudice may occur;

1. Ending unlawful use
2. Removal of invasive alien plants
3. Promoting the use of groundwater resources, where possible and available
4. Actively promoting Water Conservation and Demand Management
5 Lowering the assurance of supply without the risk of severe economic prejudice
6. Viable options for additional storage;
7. Promoting water trading

However, if this is insufficient to meet the additional demands, additional curtailments to existing lawful water users will have to be considered. In this case, it may be possible to allocate sufficient water to black applicants to start them on the path of becoming commercial water users, but not necessarily the full volume that they have requested. These users can then, once they have successfully taken up the water allocated in this way, consider trading for additional water if needed. The potential impacts on the existing lawful water users and the risks to the economy can therefore be balanced with redress needs.

Allocating water to licence applicants requires carefully balancing the impacts on existing productive water users, the need for social redress and the most economically beneficial use of water. The process must, nevertheless, give effect to the intentions of the NWA by proactively considering applications that contribute to race and gender reform. However, a carefully considered approach is important to minimise the potential for legal challenges, and to ensure that the compulsory licensing process remains consistent with the intentions of the Constitution.

5.5 Conclusions

Transforming legal access to water to redress social inequities and to promote the most economically efficient use of water poses both significant opportunities but also enormous challenges. Compulsory licensing in effect allows South Africa to "re-engineer" a new water use dispensation, reflecting not only race and gender shifts in water use, but also the most beneficial use of water in the national interest. This includes a commitment to race and gender reform, as well as to economic growth, a diverse and a stable economy, investor confidence as well as the most efficient use of water and environmental sustainability. But reallocating water to realise all these objectives holds risks. Curtailing users who are already making productive use of water, and who are contributing to the economy, not only runs the risk that existing enterprises may fail with reduced water allocations, but also that the newly established water using enterprises may struggle to establish. This is particularly a problem where larger enterprises are replaced by many smaller water users.

Moreover, water allocation reform can realise race and gender targets, but do little to address poverty – by allocating larger volumes of water to fewer large black or women owned enterprises. Conversely, widespread reallocations to large numbers of the rural poor may make better inroads into reducing poverty, but may compromise a diverse and hence stable economy. On the other hand, redress is important for social and political stability, which is vital for South Africa's future. Across Africa political instability and conflict have driven capital flight and slowed inward investment. South Africa must therefore engage the risks associated with redress not only on moral grounds but also to ensure continued economic growth.

Reconciliation of water demands and availability within compulsory licensing can be achieved both within a "*distributive model*," which can make small amounts of water available to large numbers of previously disadvantaged South Africans, and/or in a "*growth model*" by encouraging licence applications for larger scale productive uses from black and women owned enterprises. Clearly a balance between all these approaches will be required in any catchment. This will be informed not only by the requirements of the legislation and the characteristics of the catchment, but also the number and type of applications received from previously disadvantaged South Africans, and the impacts of curtailments on existing lawful water users. Changing political emphasis on redress initiatives, as well as plans for local economic development, the demographics of the catchment, and the opportunities for establishing viable small-scale businesses may also influence this balance.

South Africa should note that water use in many other African countries is dominated by large numbers of small-scale users – who are making valuable contributions to the economies of these countries. However, the GDP of these counties is also dominated by agriculture, unlike South Africa, which has a relatively more important industrial sector. This provides opportunities to cope with water stress by pushing more water into the higher value industrial uses. How the predominantly agricultural water use systems of other African countries might cope with further economic growth and increasing climate variability remains a significant challenge.

Some of this challenge may need to be addressed by reallocating between sectors and users, particularly where water is already limiting.

South Africa's NWA is now more than 10 years old. In spite of this, compulsory licensing – one of the core components of the Act – is only just being implemented. The deeply political nature of this process, the uncertainties of how reallocation will affect existing economies and the risks that protagonists are prepared to take in this respect, as well as the contribution new enterprises could make to the economy, the contribution equity can make to social and political stability, and core economic ideologies for growth and poverty reduction all contribute to make this an extremely vexing problem. The inability to balance these has certainly contributed to the lack of progress. Nevertheless, workable compromises are likely to be possible once the process is initiated.

The key lessons for reallocation processes, which have emerged from South Africa's development of the approaches to water reallocation are outlined briefly below.

Reallocation processes should be paralleled with approaches to support the uptake of basic livelihoods water use– such as by rainwater harvesting, so that basic poverty needs can be addressed without needing to curtail successful commercial enterprises to a point where their economic viability may be compromised. Beyond this, realising social redress through reallocating commercial water use should balance the risks and benefits associated with the establishment of small water using businesses with the risks and benefits of the establishing larger enterprises – which might contribute to wider developmental goals.

Water use entitlements to smallholder irrigators should minimise the administrative burden on both the water allocation authorities, and the users themselves.

Water reallocations should only be made where the risks to successful establishment of new schemes are low, and should allow for gradual uptake of the water paralleled with gradual reductions from the existing users.

Beyond this, applications from enterprises that contribute to key developmental objectives should be encouraged, especially where support to the establishment of viable enterprises will be supported from multiple government sectors, as well as the private sector.

References

Calder IR (2005) Blue revolution – Integrated land and water resource management, 2nd edn. Earthscan, London

CDE (2005) Land reform in South Africa, A 21st century perspective. Centre for Development and Enterprise Research paper No. 14, Johannesburg, South Africa

Chancellor F, Shepard D, Upton M (2003) Towards sustainable smallholder irrigated businesses (SIBU). Department for International Development, KAR Research Project, R 7810, London, UK

De Lange (n.d.) Personal communication. Private consultant to Department of Water Affairs and Forestry, Pretoria, South Africa

DPLG (Department of Provincial & Local Government) (2004) Ten year review. Department of Provincial and Local Government, Pretoria, South Africa

DWAF (Department of Water Affairs & Forestry) (1997) Overview of the water resources availability and utilisation in South Africa. Department of Water Affairs and Forestry, Pretoria

DWAF (Department of Water Affairs & Forestry) (2004) Introduction to the Orange River Basin. [Electronic version]. Retrieved June 7, 2008 from http//www.dwaf.gov.za. Department of Water Affairs & Forestry, Pretoria

DWAF (Department of Water Affairs & Forestry) (2006a) Guidelines for water allocation in South Africa. A framework for water allocation reform. Department of Water Affairs and Forestry, Pretoria

DWAF (Department of Water Affairs & Forestry) (2006b) A toolkit for water allocation reform. A manual to achieve race and gender equity in water allocations. Department of Water Affairs and Forestry, Pretoria

Falkenmark M (1995, November) Coping with water scarcity under rapid population growth. Paper presented at the Conference of SADC Ministers, Pretoria

Falkenmark M, Rockström J (2004) Balancing water for humans and nature, the new approach in ecohydrology. Earthscan, London

Kabat P, van Schaik H (2003) Climate changes the water rules. How water managers can cope with today's climate variability and tomorrow's climate change. Paper presented at the Dialogue on Water and Climate, Netherlands

Kassier Committee (1992) Report of the committee of inquiry into the Marketing Act. Department of Agriculture, Pretoria

Mayson D (2003) Evaluating land and agrarian reform in South Africa – Joint Ventures. PLAAS Occasional Paper Series No. 7, University of Western Cape

Perkins J (n.d.) Personal communication. Deputy chief engineer, KZN Region. Department of Water Affairs and Forestry, Durban, South Africa

Poonyth D, Hassan R, Kirsten J, Calcaterra M (2001) Is agricultural sector growth a precondition for economic growth? The case of South Africa. Working Paper. Department of Agricultural Economics, University of Pretoria

RSA (1996) The Constitution of the republic of South Africa (Act No. 108 of 1996). Government Printer, Pretoria

RSA (1998) National Water Act. Government Gazette Vol.398. 26 August 1998. Office of the President, Cape Town

RSA (2003) Strategic framework for water services. Government Printer, Pretoria, South Africa

RSA (2005) The National Water Resource Strategy for South Africa. Department of Water Affairs & Forestry, 2005. Government Printer, Pretoria

South African Info (2006) The Economy can grow much faster. Retrieved January 23, 2006 from http://www.southafrica.info/

Stein R (2002) Water sector reforms in southern Africa: some case studies. In: Turton AR, Henwood R (eds) Hydro politics in the developing world: a Southern African perspective, Chapter 9. African Water Issues Research Unit (AWIRU), Pretoria, pp 113–123

Van der Merwe F (n.d.) Personal communication. Deputy director, Water allocation and in stream use. Department of Water Affairs and Forestry, Pretoria, South Africa

Van der Schyff E (2004) The nationalisation of water rights: deprivation or expropriation? A South African perspective. Department of Law, Potchefstroom University for Christian Higher Education, Potchefstroom, South Africa

Van Zyl J, Binswanger H, Thirtle C (1995) The relationship between farm size and efficiency in South African agriculture, The World Bank, Agriculture and Natural Resources Department. Policy Research Working Paper 1548, Washington, USA

WRC (1996) Policy Proposal for irrigated agriculture in South Africa. Discussion document prepared by: Backeberg GR, Bembridge TJ, Bennie ATP, Groenewald JA, Hammes PS, Pullen RA, Thompson H. Report No KV96/96, WRC Pretoria

WRC (1999) Agrimarket survey. Market Surveys and Statistical Analysis Consultants, WRC Pretoria

Chapter 6
Protecting Aquatic Ecosystem Health for Sustainable Use

Harrison Pienaar, Antonia Belcher, and Dana F. Grobler

Abstract A global and South African context is provided for the policy and legislative requirements to ensure water ecosystem protection and sustainable use of water resources.

The South African resource directed measures (RDM) strategy is analysed in terms of the methods, components and their application in the implementation of provisions of the country's water legislation. The analyses include the determination and implementation of the ecological Reserve, water resource classification and the setting of resource quality objectives. It is concluded that the success of the implementation of water resource protection strategies can only be evaluated in term of the degree to which they are achieving the sustainable use and protection of water resources and their associated ecosystems. Monitoring, implementation and enforcement is the next challenge that is confronting South Africa in order to build on the successful method and policy developments that have taken place during the past decade.

Keywords Ecological Reserve • Environmental flow requirements • Water resource classification • Sustainable water use • Water resource protection • Groundwater

H. Pienaar (✉)
Department of Water Affairs, South Africa
e-mail: qin@dwa.gov.za

A. Belcher
PO Box 195, Stellenbosch 7599, South Africa
e-mail: toni.b@iburst.co.za

D. Grobler
Blue Science, PO Box 14443, Lyttelton 0140, South Africa
e-mail: dana@bluescience.co.za

B. Schreiner and R. Hassan (eds.), *Transforming Water Management in South Africa: Designing and Implementing a New Policy Framework*, Global Issues in Water Policy 2, DOI 10.1007/978-90-481-9367-7_6, © Springer Science+Business Media B.V. 2011

6.1 The Legal and Policy Context for the Protection of Aquatic Ecosystem Health

6.1.1 The Global Context

Historically, water resource management focused on the development of water resources to their fullest extent for the benefit of agriculture, industry, mining, and urban users. Ultimately, however, water is a limited resource, and the tension between a limited water resource and the need for economic growth demanded a change from a supply-side focus to a more ecologically sustainable management of water resources. This, together with worldwide changes in attitude regarding social and environmental issues, resulted in a global shift in policy and legislation relating to the management of natural resources, with much more emphasis being placed on the protection of resources.

At the United Nations Conference on Environment and Development held in Rio de Janeiro, Brazil, in 1992, Agenda 21 was adopted by more than 178 governments. Chapter 3 of Agenda 21 proposes an integrated approach to poverty relief, via community and stakeholder participation, and sustainable resource management and development, while Chapter 18 deals with the protection of the quality and supply of freshwater resources, and the application of integrated approaches to the development, management and use of water resources. Agenda 21, as a global initiative, thus informs South Africa's own approach (DWAF 2003a).

6.1.2 The National Context

In South Africa, the realities of a limited water resource were evident as early as 1956, where it was predicted that the availability of water in many parts of the country could become a limiting factor for economic development in the future. Decision-makers were also beginning to recognise the negative impacts of pollution arising from industry and urban areas. As a result, new provisions were included in the 1956 Water Act regarding the need to treat wastewater to minimum standards in order to address growing pollution problems, and the return of treated wastewater to its source so that it could be available for re-use.

During the 1980s, demands for water for domestic water supply and sanitation and to meet the needs of a growing industrial and mining sector increased and water use requirements shifted towards bulk domestic and industrial supply as opposed to irrigation. National water quality monitoring data indicated a long-term trend of degradation in many primary water resources, despite implementation of the General and Special Effluent Standards. In addition, during the late 1980s, the results of ecological research and limited monitoring showed the potentially damaging effects of over-abstraction and the regulation of rivers on aquatic ecosystems. The global trend towards recognition and incorporation of environmental concerns into water resource management added pressure for change (WWF 2003).

In 1996, the Constitution of the Republic of South Africa (Act No 108 of 1996) laid the foundation for a democratic society in which government is based on the will of the people. There is also a promise, by government, to improve the quality of life for all citizens. According to Chapter 2 of the Bill of Rights, everyone has the right to an environment that is not harmful to their health and well-being, to have an environment that is protected for the benefit of present and future generations, and to have access to sufficient food and water. These statements set the scene for the development of policies and legislation that are socially and economically enabling, while also ensuring sustainability.

In response, South African policy has focused strongly on equitable and sustainable social and economic development for the benefit of all South Africa's people. Many existing laws, however, including water legislation, were not focused towards achieving these objectives. The National Water Policy for South Africa, adopted by Cabinet in 1997, was introduced in response to the new direction set by government and as part of a thorough review of existing water law. This policy was founded on and guided by the Water Law Principles. Three fundamental objectives for managing South Africa's water resources arise from the Principles: equity, efficiency and sustainability.

The resulting 1997 White Paper on a National Water Policy and the National Water Act (NWA) of 1998, were seen as key milestones in the process of water law reform with far-reaching effects on social, economic and environmental issues in South Africa (De Coning and Sherwill 2004). The NWA is the principal legal instrument relating to water resources management in South Africa and contains comprehensive provisions for the protection, use, development, conservation, management and control of South Africa's water resources.

However, since water is essential for all life and most human activities, there are many other policies and laws, administered by a number of departments, that directly or indirectly affect water resources. These include the Strategic Framework for Water Services (RSA 2003), the Water Services Act (No. 108 of 1997), the National Environmental Management Act (No. 107 of 1998), and the Environment Conservation Act (No. 73 of 1989). Successful water resources management therefore, depends on co-operation between government, water institutions and other organisations, water users and other stakeholders.

6.2 Characteristics of South Africa's Aquatic Ecosystems

Different combinations of climatic, geological and topographical conditions at different latitudes have created a variety of aquatic ecosystems in South Africa. Surface freshwater bodies are either running (lotic) water systems such as rivers and open-mouthed estuaries, or standing (lentic) water systems such as lakes, wetlands and closed-mouth estuaries (Davies and Day 1998). Rivers in some areas flow most strongly in winter, whereas in other areas they flow strongly in summer.

Many rivers only have seasonal flow, while some flow only briefly after rains. A few lakes and wetlands remain inundated for the entire year, while others dry up for part of the year. Water chemistry also differs from region to region.

The NWA recognises these complexities and the need to manage a water resource in its entirety as an aquatic ecosystem. This includes the physical or structural habitats, both in-stream and riparian, the water, the aquatic biota, and the physical, chemical and ecological processes that link them. This new, wider understanding of the various components of a water resource was a major shift from the past approach to water resource management, which tended to look solely at the quantity and quality aspects of water and its importance in terms of use.

The NWA defines water resources as watercourses, surface water, estuaries and aquifers, where a watercourse may be a river or spring, a natural water channel, or a wetland, lake or dam. Wetlands are seen in the NWA as "land which is transitional between terrestrial and aquatic systems where the water table is usually at or near the surface, or the land is periodically covered with shallow water or would support vegetation typically adapted to life in saturated soils". A wetland is, therefore, defined in terms of its hydrology, soils and the associated plant life. The NWA defines an estuary as "a partially or fully enclosed body of water - *(a)* which is open to the sea permanently or periodically; and *(b)* within which the sea water can be diluted, to an extent that is measurable, with freshwater drained from land". Estuary characteristics are determined by river inflow patterns and quality, hydrodynamics and sediment dynamics, and the influence of the sea (DWAF 2002). Both wetlands and estuaries are also managed under legislation implemented by the Department of Environmental Affairs. This has caused delays in the development of resource protection measures for wetlands and estuaries due to a need to reach consensus between government departments.

Groundwater has been addressed in the NWA through the definition of an aquifer as "a geological formation which has structures or textures that hold water or permit appreciable water movement through them". During wet seasons, surface water contributes to the recharge of groundwater systems, while during the dry seasons, groundwater plays a critical role in sustaining the ecological functioning of surface water systems. Much of South Africa's groundwater (90%) occurs in hard rock and has a highly variable nature (DWAF 2004a). Groundwater, prior to the promulgation of the NWA, was considered private water and received virtually no protection, except in the so-called Government Subterranean Water Control Areas. The change of groundwater from being private water to being considered part of the national water resource managed under public trust by the Minister of Water Affairs and Forestry has required a change in the management of this water resource, where protection is a vital component.

Due to the different characteristics of water resources, they fulfil a variety of functions and offer a range of ecological goods, services and attributes (EGSAs) to both the environment and society. These can include

- Supply of primary resources (basic human uses – cooking, cleaning and washing, supply of reeds, clay and fish)
- Regulatory functions (climate, nutrient cycling, erosion and sedimentation)

- Recreational/aesthetic value (fishing, boating, swimming and appreciation of nature)
- Economic value and functions (water supply to farming and industry, breakdown of pollutants, flood attenuation)
- Cultural value (baptism); and
- Ecological functions (provide habitat for all forms of life, part of a greater ecosystem and food web, biodiversity importance)

Managing water resources for their diversity also means managing those resources for the diversity of goods, services and attributes that they provide (Palmer et al. 2002).

Freshwater systems are, also, complex ecological systems. For example, in riverine ecosystems, the different magnitudes of flow that form part of the natural flow regime affect the system in a variety of ways:

- Higher flows define the geomorphologic character, shape and size of a river channel, as well as the size of the bed particles, which generally decrease in size downstream where boulders give way to cobbles, gravel, sand and finally, mud.
- Baseflow, often contributed to largely from groundwater, maintains the basic ephemeral, seasonal or perennial nature of the river.
- Small pulses of elevated flow, or freshes, stimulate spawning in serial-spawning fish species, flush out poor quality water, mobilise sandy sediments, and create the flow variability that seems to be such an important influence on the character and functioning of riverine ecosystems in arid areas.
- Floods reset a wide spectrum of conditions in the river, triggering and synchronising activities as varied as the upstream migration of fish and the germination of riparian seedlings. Floods also scour riverbeds and estuaries, ensuring, amongst other things, the accessibility of nursery areas within the estuaries to marine fish, as well as the maintenance of habitat diversity (Brown 1996).

Understanding each individual aquatic ecosystem, the components that make up that ecosystem, as well as their intra- and inter-relationships, is critical to managing water resources and ensuring their protection.

Quantification of the water requirements (both quantity and quality) for aquatic ecosystems is based on an understanding of the functioning of ecosystems and their habitat requirements. The quantification is also, however, based on hydrological data for the water resources. Because of the high diversity in rainfall, runoff, and thus stream flow, both spatially and temporally, low flows and floods are also highly variable in duration and severity in space and time. All of this variability and complexity makes the prediction of Environmental Water Requirements (EWR) extremely difficult.

The highly variable nature of river systems in South Africa, and the resulting variable water availability, has brought about the need for water resources to be developed in order to increase the assurance of supply. Currently, most of our water requirements are dependent on surface water resources. The total mean annual runoff (MAR) for the country, under natural (undeveloped) conditions, is estimated at close to 50 billion m^3/a, with a further 6 billion m^3/a available from groundwater. Depending on the level of protection required for Ecological Goods, Services and

Attributes, a portion of this runoff needs to remain in water resources to satisfy the requirements for the ecological water requirements (the Reserve), while only part of the remainder can, practically and economically, be developed for use.

In general, surface water resources have been highly developed over most of the country, through the construction of about 320 major dams of more than 1 million m^3 in size. This equates to about 66% of the total MAR of surface water. Groundwater utilisation, on the other hand, has been limited mostly to the rural and more arid areas, due to the geology of the country, much of which is hard rock. Large, porous aquifers occur only in a few areas (DWAF 2004). In many cases, particularly in the drier, rural areas, groundwater is the sole source of water supply to many towns and communities. Where there is such a high level of dependency on groundwater for water, the need to manage and protect those resources becomes all the more critical. Currently, approximately 21 billion m^3/a of water is utilised – of this, 52% is for agriculture, 8% for forestry, 11% for industry, 10% for domestic use and approximately 19% is set aside for ensuring a sustainably-functioning environment (Minister of Water Affairs and Forestry 2006).

Many of our river systems are already stressed, with no more water being available for allocation or with water only being available for use during the wet seasons. In these catchments, the environmental water requirements are not being provided for in terms of flow volumes, as well as the timing and variability required. Management of these resources is becoming increasingly difficult, especially where new water users need to be catered for and where the imbalances of water allocation in the past needs to be rectified. In these catchments, trading of water-use licences and water allocation reform management options are being followed. Some level of protection of aquatic ecological functioning in these catchments is critical.

6.3 Water Resource Protection in the Context of Integrated Water Resource Management

Since South Africa's water is scarce and unevenly distributed throughout the country, it is one of the limiting factors to economic production and growth – in short, society cannot sustain any economic activity without the goods and services provided by water resources. For these reasons, an integrated water resource management approach is required that promotes the co-ordinated development and management of water resources to maximise social and economic development, while ensuring equitable and sustainable utilisation of the resource.

If properly managed, water resources can be utilised indefinitely, such that people can benefit from their innate ability to recover and reset themselves over time. Exploitation of land and water resources without understanding their limits, however, can affect catchment processes and functions, and have a negative impact on the ecological communities that have evolved together, in a balance, over millions of years. When the balance, resilience, integrity and ecological health of water resources are disturbed, they can become dominated by one or a few hardy organisms, such as blackfly, algae, or introduced pests such as water hyacinth, which have no local natural enemies. Alternatively, when water resources are over-utilised, rivers that were permanent may stop flowing or wetlands may dry up. If such levels of exploitation are allowed, the rights of downstream and other users are impacted on.

This doesn't mean that water resources should not be utilised. It just means that the natural components, processes and functions, on which the water resources depend, should be protected and taken into consideration when decisions are made concerning the use and development of water resources. Responsible management is the core of sustainable utilisation of the water resource and is central to integrated water resource management. In order to understand and manage the water resources within their sustainable limits, a Resource Directed Measures strategy was developed and is used to give effect to the provisions of Chapter 3 of the National Water Act. This set of measures consists of a procedure for classifying water resources to achieve a desired level of protection, as well as the water quality and quantity requirements and the resource quality objectives that must be met in order to achieve the desired ecological state within the water resource so as to sustain a certain level of ecosystem functioning, and to provide for the desired goods and services.

The Resource Directed Measures have also been developed in such a manner that they balance social and economic needs with the environmental needs of water resources. The degree to which a water resource can be utilised, as well as the quantity and quality of water set aside for the environment, is determined by the water resource class selected. The procedure developed for the selection of the water resource class is specifically designed to achieve long term sustainable use by considering the social, economic and environmental aspects of the resource.

6.4 Resource Directed Measures (RDM)

RDM is a water resource management strategy with the objective of ensuring sustainable utilisation of water resources. The strategy comprises the Classification of water resources (see Section 6.10), the Reserve and Resource Quality Objectives (see Section 6.11), and specifically addresses the determination of the level of protection that should be afforded to a water resource to ensure that it continues to function at in desired ecological state. Another strategy – the Source Directed Control (SDC) strategy – exists for managing the impact of land-use activities on water resources. RDM should be implemented in conjunction with the SDC strategy to ensure that an integrated approach is taken that looks at the social, economic and environmental requirements. Currently, however, decisions regarding water resource utilisation are generally not integrated, and RDM and SDC are applied in isolation.

6.4.1 The Ecological Reserve: From Conceptualisation Towards Implementation

During the late 1980s, the policy relating to water resource management in South Africa shifted from not recognising the value of aquatic ecosystems to the view that aquatic ecosystems had legitimate water requirements. but were competing users of water, and could be allocated water provided this did not compromise other water users to any great degree and did not constrain economic development (WWF 2003).

The post-1994 review of the water law, however, provided opportunity to place important policy objectives for the sustainable use of aquatic ecosystems on the agenda (De Coning and Sherwill 2004). These objectives were formalised in the 'Reserve' concept, embedded in both the founding Principles of the Act and Chapter 3 of the NWA itself, where the Reserve is defined as that quantity and quality of water required –

(i) To satisfy basic human needs for all people who are, or who may be, supplied from the relevant water resource, (Basic Human Needs Reserve).
(ii) To protect aquatic ecosystems in order to secure ecologically sustainable development and use of the relevant water resource' (Ecological Reserve – quality and quantity).

In this regard, South Africa was the first country to formally legislate for environmental water requirements.

6.4.2 RDM Methodology Development

To meet the requirements of the NWA, existing resource protection methodologies needed to be refined and new methodologies developed to determine the Reserve for both quantity and quality, as well as for the various components of aquatic systems, i.e. rivers, wetlands, estuaries and groundwater. Fortunately, since the 1980s, a considerable amount of effort within the South African scientific community had been focused on developing environmental flow assessment methods as part of environmental impact assessments for water resource developments. These methods focused mostly on rivers, to determine the In-stream Flow Requirements (IFR) and on estuaries, to determine the Estuarine Flow Requirements (EFR).

A selection of methods was, therefore, available when the drafting of the new water legislation commenced in 1997. The methods had not been specifically tailored to the requirements as laid out in the proposed legislation, however, and the existing scientific approaches and procedures provided the foundation for a new suite of methodologies (Louw et al. 2002). These methods are still at various stages of development, from being well-developed for the determination of

Environmental Water Requirements (EWRs) in rivers, to requiring some refinement for determining the requirements for wetlands. The Building Block Methodology (BBM) formed the foundation for later, more advanced methodologies such as DRIFT and Stressor Response.

Also addressed during the development of methodologies for Reserve determinations, was the level of assessment required. The levels varied from an estimate of environmental flows required at a broad scale for planning purposes, to detailed requirements specified to inform operational rules for large water resource developments. Four different levels of ecological Reserve determination were identified and, initially, described in terms of the time it took to complete them. These ranged from a Desktop method, which could be undertaken in a matter of hours, through to a Comprehensive method that was estimated to take approximately 2 years. It was assumed that the degree of confidence in the results of an assessment would increase in direct proportion to the time and cost involved. However, in practice, it has been found that this is not necessarily the case. Any determination that does not satisfactorily define the biophysical relationships between ecological functioning, channel hydraulics, geomorphology, the hydrological regime, and water quantity will return information that is of low confidence, no matter how much cost and time was spent on the study. Thus, "Desktop", "Rapid", "Intermediate" or "Comprehensive," describe the Reserve determination method, while the terms "low", "medium" or "high", refer to the level of confidence, reflecting the integrity of the results (DWAF 2003a).

The different methods of Reserve determination also vary in terms of their ability to allow for public participation during the study. Initially, the Comprehensive Reserve determination allowed for a public participation process and an assessment was undertaken as part of these higher level studies of the socio-economic impacts of different water resource development scenarios. During the initial stages of implementation (1999–2002), it became clear that public participation was too costly and time-consuming to take place during the Reserve determination study and that it should actually take place during the classification of water resources when all social, economic and environmental aspects and their implications could be taken into consideration.

6.4.3 *Ecological Reserve Determination Methodologies for Rivers: Quantity*

Environmental flow requirements for rivers have been defined as that water left in a river system, or released into it, for the specific purpose of managing the condition of that ecosystem (Brown and Joubert 2003). Internationally, during the last 5 decades, about 100 different approaches have been described for advising on environmental flows worldwide, and more than 30 countries have begun to use such assessments in the management of water resources (Tharme 2002).

The currently-accepted method for comprehensive determination of the water quantity component of the ecological Reserve for rivers in South Africa is the Building Block Method (BBM). Using BBM with the best available knowledge and expertise, the assessments are designed to produce a recommended, modified flow regime for the targeted river that will facilitate its maintenance in some pre-determined, desired future condition. The development of BBM was funded by the Water Research Commission (WRC) and has been tested and adapted in a number of Intermediate and Comprehensive Reserve determinations. The methodology represents the minimum procedural requirements necessary to achieve the appropriate level of confidence in the determination of the water quantity component of the Reserve for a river reach under consideration. It is probably the most frequently-applied, holistic environment flow method applied throughout the world thus far, where many applications of BBM, or adaptations thereof, have been undertaken in Southern Africa and Australia (Tharme 2002).

A methodology that recently evolved from the BBM is the Downstream Response to Imposed Flow Transformations (DRIFT) process. It focuses on the identification, by a multidisciplinary specialist team, of the consequences of reducing discharges through a series of flow bands associated with particular sets of biophysical functions, and of specific hydrological and hydraulic character, in terms of deterioration in system condition. As the methodology is scenario-based, there is considerable scope for the comparative evaluation of the consequences of a number of recommended flow regimes (Tharme 2002, Brown et al. 2006). This method has been applied with much success to a number of Intermediate and Comprehensive Reserve determinations in South Africa, as well as internationally.

While much attention has gone into the further development of the methodologies utilised for determining the EWRs for rivers since the promulgation of the NWA, little attention has been paid to assessing the validity of these assessments. No fully-successful implementation of these flow assessments or Reserves has taken place to such an extent that the results can provide for the verification of the methods. Implementation of the Reserve implies giving effect to the Reserve by achieving a desired level of ecological functioning. This needs to happen through

- Ensuring that the required EWRs are being met (This could be achieved in a number of ways, such as the refusal of water-use licence applications due to insufficient water being available to meet EWR, the incorporation of conditions in ad hoc water-use licences to address the EWRs, or the incorporation of environmental flow releases into the operational rules for large in-stream dams)
- Monitoring compliance with water-use licence conditions that relate to the Reserve, as well as monitoring downstream flows compliance against EWRs and RQOs; and
- Assessing whether the recommended EWR achieved what they were set out to do in terms of the given eco-specifications

Desktop and Rapid methods for determining the quantity component of the ecological Reserve, on which many water allocation decisions are made, are based on extrapolations of previous high confidence Reserve determinations. To improve the confidence of these Reserves, it is critical that the recommendations of previous high confidence Reserve determinations be assessed against monitoring results. Implementation, monitoring and testing are, therefore, urgent requirements for ensuring the successful implementation of the Reserve.

6.4.4 Ecological Reserve Determination Methodologies for Rivers: Quality

The task of an ecological Reserve assessment is to provide both quantified and descriptive information about the pattern and reliability of environmental flows, with information on flow frequency, magnitude, and duration, so that an entire modified flow regime can be provided. However, methods for quantifying environmental water quality still focus mainly on only magnitude (concentration). Frequency and duration are only taken into account via flow-concentration modelling. The ecological Reserve for water quality is provided as class boundary-value concentrations for each variable (Palmer et al. 2002).

The constituents and approach followed for the determination of the water quality Reserve still receive much debate. From the many possible water quality variables, those considered in South Africa include: inorganic salts (sodium chloride, sodium sulphate, magnesium chloride, magnesium sulphate, calcium chloride, calcium sulphate); nutrients (phosphate as $PO4^{-3}$, and total inorganic nitrogen); physical variables (turbidity, pH, oxygen, and temperature); and those toxic substances listed in the South African Water Quality Guidelines for Aquatic Ecosystems (DWAF 1996b). In addition, bio assessments such the aquatic invertebrate index SASS, algal abundance and toxicity tests are utilised for supporting information (DWAF 2001a).

The final water quality report provides qualitative and quantitative water quality objectives for each resource unit, at a range of ecosystem health classes (Palmer et al. 2002). The water quality requirements are linked to the complexity of ecosystem structure and function through components that assess habitat (hydraulic and geomorphologic), biota (fish, invertebrates, vegetation – riparian and in-stream), and the responses of biota to the stress of altered flow and chemical variables.

The development of methods for environmental water quality assessments, and the integration of water quality with flow in protective water resource management still requires much work and, more importantly, testing. A particular problem with the current methodology is the availability of data for those constituents required to conduct the assessment. Implementation of the requirements of the quality component is also problematic and, in practice, the quality requirements of downstream water users are often more sensitive than the aquatic ecosystem.

6.4.5 Ecological Reserve Determination Methodologies for Estuaries: Quantity and Quality

Estuarine Freshwater Requirement studies by estuarine scientists and managers, prior to promulgation of the NWA, were mostly conducted for DWAF to inform large water resource developments. These studies provided the foundation for the development of the estuarine Reserve determination methodology that has since been formalised and documented (DWAF 1999c).

Current available information on local estuaries is highly variable, ranging from almost no data to good data and information. Estuarine Reserve determinations, therefore, often require specific additional fieldwork in order to meet the data requirements required by the methodology (Adams et al. 2002). Where data has been limited, particularly with regard to hydrology, appropriate levels of confidence are affected. The costs and time required for Reserve determinations are relatively high where there is very little data available. Currently, methodologies are being developed for Rapid and Desktop Reserve determination methods for estuaries. Due to the current constraints of data availability, few estuary Reserves have been undertaken as the time required and costs related to these assessments are prohibitive.

In general, the estuarine freshwater requirement is a relatively high proportion of the natural freshwater contribution to the estuary, specifically in relation to the environmental flow requirement for the upstream river reach. This is because an estuary requires a large flood component for the scouring of the estuary and to maintain the hydrodynamics of its mouth, while low flows are essential for maintaining the salinity profiles within the estuary that provide for its productivity. These high-flow requirements are usually viewed as limiting water resource developments within the catchment. The output from an estuarine Reserve determination is given in terms of the development scenarios that can be considered in the catchment of that estuary that will not result in an unacceptable deterioration of the estuarine ecosystem. These flow requirements usually relate to the volume of flow that needs to enter the estuary, and not the variability and timing of that inflow. This should provide water resource managers with more options to allow for water resource development, while taking into account the estuarine flow requirements.

6.4.6 Ecological Reserve Determination Methodologies for Wetlands: Quantity and Quality

The current wetland RDM method only covers the following classes of South African wetlands:

- Lacustrine: Limnetic and Littoral (natural freshwater lakes)
- Palustrine: Flat, Slope, Valley Bottom, Floodplain (freshwater marshes, peatlands, springs, swamp forest, floodplains); and
- Endorheic (permanent and seasonal pans) (DWAF 1999b)

Due to the relatively new and untested status of the Reserve methodology for wetlands, no Rapid or Desktop methods exist for determining the wetland component of the Reserve. The methods for determining the EWR of wetlands are, currently, under further development. Several of the existing preliminary wetlands Reserve determination tools and approaches constitute modifications of similar protocols developed for rivers.

The output from a wetland reserve is often simply the water levels required for inundation of the wetland over a period, and at a certain time of year. A number of the recommendations, from the Reserve determinations, are usually non-flow related. These aspects are usually easier to manage than the flow requirements. Thus, an important aspect in the implementation of the wetland Reserve component, is understanding the interlinkages between the wetland and the groundwater, river and estuary components of the Reserve, to be able to manage inflows and outflows from the wetland to conform with the wetland Reserve requirements.

6.4.7 Resource Directed Measures for Groundwater: Quantity and Quality

Initially, the generic process for surface water Reserve determinations was also proposed for the determination of the groundwater Reserve (DWAF 1999d). Recent revisions of the groundwater methodologies have, however, proposed different approaches. While most revisions follow the same basic approach, the steps within that approach are not uniform.

Implementation of the groundwater Reserve is still, however, problematic. The Reserve for groundwater is considered as only that part which sustains basic human needs and aquatic ecosystems. As the ecological component is only that groundwater which contributes towards baseflow in surface water resources and, as groundwater is far more widespread than surface water resources, the groundwater Reserve alone cannot afford adequate protection of aquifers and needs to be implemented in conjunction with RQOs which afford greater protection to the aquifers. In addition, the interpretation of the ‘Groundwater Reserve’ and the integration with surface flows is poorly defined and understood and difficult to translate into water use management actions.

6.5 The Basic Human Needs Reserve

The Basic Human Needs (BHN) Reserve is intended to provide for the essential needs of those individuals reliant on a water resource either for water for basic human needs. This is the first right to water in the NWA and is intended to ensure that sufficient water remains in water resources to provide for those communities still dependent on water resources for their water supply. The water set aside for this is calculated at 25 L/person/day in low confidence Reserve determinations. For higher confidence Reserve determinations, a socio-economic assessment is undertaken to evaluate the dependence of communities on the water resources, not only for basic water supply but also in terms of providing for sustainable livelihoods, in which case the allocation is adjusted to 60 L/person/day.

6.6 Reserve Requirements for Planning Purposes

During 1999, DWAF compiled and calibrated a Water Situation Assessment Model (WSAM), to determine water availability in South Africa, for planning purposes. Ecological Reserve requirements were needed as an input to the model and DWAF initiated a national compilation of ecological data for the river mainstreams in all 1946 quaternary catchments. Ecologists compiled a summary table of the ecological importance and sensitivity (EISC) and present ecological status (PESC) for each quaternary catchment. This data was utilised to assign preliminary classes to and calculate the Reserve requirements for each of the quaternary catchments, by means of the ResDSS (Reserve Decision Support System (Hughes and Münster 2000). This Reserve determination method was referred to as the Desktop method and was used for development of the National Water Resource Strategy (NWRS) and the Internal Strategic Perspectives (ISP) (fore-runners to catchment management strategies).

The ResDSS was based on the extrapolation of the environmental flow requirements for high confidence Reserve determinations carried out mostly in the northern parts of South Africa. These Reserves were less applicable to the rivers to the south of the country, where the climate and flow regimes are very different. The ResDSS has recently been updated, based on the results from high confidence Reserves undertaken in rivers in the southern and western Cape. The ResDSS was only intended for planning purposes, as it was felt that the model was not conservative enough for more detailed decision-making. The model is, however, being utilised to consider water-use licence applications where the potential impact on the water resource is small. Implementation of the Reserve through water use licences is limited as it is not feasible to require complicated operational rules of small-scale water users. Therefore, it does not make too much sense undertaking high confidence Reserve determinations for such instances.

A general trend that was observed in the original outputs from the ResDSS, was that an unmodified river (A class) required approximately 45% of natural MAR, a slightly modified river (B class) required approximately 35%, a moderately modified river (C class) required approximately 25%, and a largely modified river (D class) required 10–20%. Due to the variability of our river systems, however, such parallels cannot be drawn across the country. In the Western Cape rivers, for instance, much higher flow requirements have been established, due to the greater contribution required from large floods that are important for the movement of large boulders and rocks in the river systems.

6.7 Reserve Determinations for Ad Hoc Water-Use Licensing

The NWA requires that the Reserve be considered prior to any water-use allocation. This requirement is applicable for the authorisation of any of the 11 water uses listed in Section 21 of the NWA, and for the transfer of water use between water users. Essentially, this allows DWAF to progressively implement provisions of the NWA, especially in instances where non-compliance with the Reserve is already occurring. The NWA also makes provision for the progressive implementation of protection measures by allowing preliminary determinations of the Reserve and Resource Class, which can later be superseded by the results from higher confidence studies.

Between October 1999 and 2008, the RDM directorate has received approximately 1,600 requests for Reserve determinations. Of these, the RDM directorate completed approximately 900 preliminary Reserve determinations, 750 of which were approved. The approved preliminary Reserve determinations vary from low confidence to high confidence. "Preliminary" refers to the legal status of the determined Reserve i.e. it has been determined prior to the establishment and implementation of a classification system. With time, the intention is to replace the low confidence Reserve determinations with high confidence determinations, and to replace preliminary Reserves with final Reserves. These Reserves will still be required to be assessed over time and modified to ensure that they meet with the objectives that they were set to meet.

During this period, a number of practicalities regarding Reserve determinations and implementation became apparent. For instance, the relevance of Reserve determinations for water uses relating to the alteration of the bed, banks or characteristics of a watercourse, the impedance or diversion of flow, or for recreational use, has often been questioned, as the impact of these water uses is more upon the habitat and biota than on the water quantity and quality. The revision of the General Authorisations in 2003 provided an opportunity for a number of these water-use activities that are considered to have a low impact on water resources and, more specifically, on the Reserve requirement, to be generally authorised which does away with the need for the Reserve to be determined. Unfortunately, General Authorisations do not distinguish between sensitive aquatic ecosystems from those in the same catchment that are very resilient or have a very low ecological importance.

The translation of the ecological Reserve into a format that can be utilised by the water resource manager remains a challenge, as this requires the calculation of the water balance, sometimes at a small sub-quaternary catchment scale, before decisions can be taken.

6.8 High Confidence Reserve Determination Studies

A number of high confidence Reserve determination studies have been conducted since 1999, utilising Comprehensive and Intermediate Reserve determination methodologies. These results will remain valid, and will be used during the process of water resource classification.

Nearly all of the above comprehensive Reserve determinations were undertaken in catchments where there is already an imbalance between water-use requirements and the available water for use. Implementation of the Reserve in such catchments is not simple and requires an integrated approach to ensuring that the social, economic and environmental water-needs of the catchment are met.

The solutions to the conflicts around water in water-stressed catchments require decision-making by water resource managers, and cannot be resolved through Reserve determinations. Often, the Reserve determination study has been seen as delaying the decision-making process, causing frustration amongst water users who wish to use the water as soon as possible. The reality is, though, that for many of these higher confidence Reserve determinations, very little implementation of the EWR recommendations has taken place. Because of the complicated nature of the Reserve EWR results, the results of the Reserve studies need to be translated into a Reserve implementation plan before they can be implemented by water resource managers.

In addition, the inclusion of aquatic biodiversity conservation in the comprehensive determination of the EWR requirements within catchments has not been a smooth process. Current higher level Reserve determinations are starting to address these requirements.

Box 6.1 Case Study: A Comprehensive Reserve Determination for the Thukela River

In 2000, DWAF initiated a study to determine a comprehensive Reserve for the Thukela River (river, groundwater and estuary components), to provide information for the planning of two dams, and for the consideration of water-use license applications. The study also included an assessment of the BHN Reserve, an economic study and stakeholder consultation.

The study area was particularly large (29,039 km^2) and complex. During the initial stages of the study, the water quality methods were being redesigned and these developments were tested in the study. As no method for water quality for estuaries existed at the time, the methods used for rivers were adjusted to suit the estuary. The integration of water quality and quantity was addressed by providing water quality consequences for each quantity scenario provided. An approach was also developed for the integration of the quantity estuary and river results. The groundwater contribution to the Thukela River baseflows was thought to be insignificant, therefore, the groundwater component was assessed at a scoping level only. In addition, an overview of the costs and benefits, in terms of goods and services provided by the environment for each of the Reserve scenarios was generated, using existing economic information relating to the utilisation of water resources in the catchment.

(continued)

Box 6.1 (continued)

The stakeholder process dictated the latter part of the programme. Sufficient time was required between the generation of results and the stakeholder process itself, for the conversion of the results into a user-friendly format and for capacity-building to take place. Time was also needed for the assessment of stakeholder input. When all the results became available from the yield modelling and ecological assessment, time was required for the resource economic modellers to assess the results and provide economic consequences for each result.

The Thukela study provided a significant opportunity to test, refine and develop methods and approaches for incorporation into the Reserve methodologies to streamline future comprehensive Reserve assessments. The time and cost associated with the study also demonstrated that this kind of investment should only be made in rivers of major socio-economic and ecological importance.

6.9 Implementation of Reserve Requirements

DWAF has invested, substantially, in the development of methodologies for and the determination of Reserve requirements for various water resources. It remains, however, a key challenge to monitor, implement and enforce Reserve requirements and other aspects of water resources protection. Compliance with Reserve requirements needs to be enforced, *inter alia*, through licence conditions. Unfortunately, implementation is the mandate of under-capacitated regional offices of DWAF and CMAs, which are still in the process of being established.

During the late 1990s, it was estimated that very few of the 25 IFR determinations studies that had been undertaken since the early 1980s, had actually been implemented. These studies were primarily undertaken to determine the flow releases required from existing dams. The lack of implementation was mostly the result of a combination of lack of legislative mechanisms to enforce the specified releases, lack of operating rules and the over-complication of operating rules by aquatic scientists who often insisted simulating releases, as accurately as possible, with natural flow events, and lack of political will.

Historically, water was released from large in-stream dams for downstream water use, typically for irrigation. These releases would take place as a constant release throughout a period of the year when, often, the flow in the river is at its lowest, and thus, not able to provide for the required irrigational water use. It has been proposed that these releases contribute towards meeting the required EWR, but in most cases these releases do not have the flow right variability or occur at the right time of year to meet ecological reserve requirements.

For small-scale water users, it is seldom feasible to write complicated operational rules for environmental water releases into water-use licences, particularly where little hydrological data is available and the extrapolation of low confidence Reserve EWRs becomes very inaccurate. In catchment areas where basin studies have been undertaken, the confidence and accuracy of information allows more appropriate decision-making to take place. However, little or no compliance monitoring is undertaken of licence conditions, raising the question of whether Reserve release requirements are, in fact, being complied with. Equally, achieving proper environmental releases from dams, as has been specified in the Reserve studies, has been problematic, partly due to not having sufficient predictive monitoring in place to effectively mimic the natural flow pattern. In some cases there has been insufficient understanding of the Reserve requirement, particularly in terms of the variability of flows required.

Currently, in the case of new, large in-stream dam developments, detailed studies are undertaken that include the hydrology, environmental impact assessments and high confidence Reserve determinations. The output from the studies provides the necessary information to be able to develop operating rules for the release of environmental flows from the dam. The Berg River Dam development, in the upper reaches of the Berg River, near Franschoek, is an excellent example of where all the parties involved in the development have committed to giving effect to the ecological Reserve.

Resource monitoring and compliance auditing is lagging behind the ability to determine the Reserve. A balance between the human and capital resources needed to develop methodologies and determine Reserve requirements, and those needed for implementation will need to be achieved. It should also be recognised that compliance monitoring of Reserves, and the evaluation of whether recommended EWRs attain the desired ecological functioning that they were set to achieve, requires long term data collection over a period of 5 years or longer.

6.10 Water Resource Classification

A Water Resource Classification Systems (WRCS) has been developed that provides a consistent framework according to which water resources can be classified. The system provides guidance on public participation in the process of classifying water resources, and provides specifications against which management decisions can be made about the nature and extent of permissible and sustainable resource use.

The guiding principles of the WRCS are

- Balancing resource protection and use in line with societal norms and values
- Maintaining ecosystem integrity at a level that ensures the continued delivery of desired ecosystem goods, services and attributes for use
- Balancing catchment-level decisions with national interests (and, where appropriate, international obligations)

- Transparency
- Implementability and reasonable cost
- Recognition of the interdependency of the hydrological cycle and the inter-linkages between rivers, aquifers, wetlands and estuaries
- Legal defensibility and scientific robustness
- Classification at an appropriate scale
- Auditability and enforceability
- Lowest level of contestation and the highest level of legitimacy; and
- Utilisation of existing tools, data and information within a framework of consistent class determination (DWAF 2006b)

Some comments relating to the WRCS indicate potential problems with its implementation because of technical complexity and lack of capacity. The lack of economic data at an appropriate level and at a sufficient level of detail, would further hamper implementation.

The actual classification of water resources will be implemented in a phased and progressive manner, as and when required, over the next number of years. The desired characteristics of the resource will be represented by a Management Class (MC). The economic, social and ecological implications of choosing a MC will need to be established and communicated to all Interested and Affected Parties (I&AP) during the Classification Process. The MC will, therefore, outline those attributes that the custodian, DWAF, and society require of different water resources. The outcome of the Classification Process will be the approval, by the Minister or her delegated authority, of a MC (as well as the Reserve and RQOs that will give effect to the class), for every significant water resource (river, estuary, wetland and aquifer). This will be binding on all authorities or institutions when exercising any power, or performing any duty, under the NWA.

The classification system contains three management classes. The MC ranges from Natural to Heavily Used/Impacted (Table 6.1), and essentially describes the

Table 6.1 Water resource classes

Class Name	Class Description
Natural	A reference condition for other resources classified at greater levels of impact. Resources in other classes will be defined in terms of the degree of deviation from the Natural class
Moderately used/impacted	Resource conditions that are slightly, to moderately, altered from the Natural class reference conditions due to the impacts of human activity and water use
Heavily used/impacted	Resource conditions that are significantly changed from the Natural class reference conditions, due to the impacts of human activity and water use, but that are, nonetheless, ecologically sustainable
Unacceptably degraded Resources	Because of over-exploitation or major alteration to their physical structure, some resources are already in a condition that can be described as unacceptably ecologically degraded.

desired condition of the resource, and the degree to which it can be utilised. The class descriptions are given in terms of utilisation and not in terms of the originally-proposed descriptors for protection (natural, good and fair). This highlights the viewpoint set out in the national policy that aquatic ecosystems are to be protected *for use* – thus, protecting a "pristine" river, is not necessarily the objective of the National Water Act. The "Unacceptably degraded resources" class is an unsustainable condition and will only be used to describe present status. Where the present state of a resource is "Unacceptably degraded", the management class will be set at least one class higher, as heavily used/impacted with the aim to rehabilitate the resource to the desired state.

The classification of water resources forms an integral component of integrated management of water resources. Accordingly, the Classification Process cannot occur in isolation from the integrated planning of water resource protection, development and utilisation, and the management and control of water use. A key component of achieving effective water resources management is, therefore, an iterative process of evaluating water use and protection scenarios with stakeholders, where the economic, social and ecological trade-offs will be made, and out of which will emerge the MC, Reserve, RQOs, and possibly an allocation schedule that can be incorporated into the CMS.

Implementation, enforcement, and monitoring – to evaluate whether the predicted economic prosperity, social well-being and ecological condition have been achieved – will be essential to ensure long term sustainability of the selected management option. The desired level of protection will have been achieved if the conditions, appropriate to the designated class, are achieved (DWAF 2006b).

6.10.1 Specific Challenges in the WRCS and the Classification of Water Resources

Given the socio-economic challenges of South Africa, the protection of water resources, and the perception that resource protection prevents economic development, will continue to present a substantial challenge. At the same time, the environment is complicated by the decentralisation of functions to Catchment Management Agencies (CMAs). DWAF will assume custodianship of the resource by overseeing and regulating the resource and its management. The CMAs will be responsible for managing the resource. Accordingly, the CMAs may develop recommendations on the class, which will be assessed and reviewed by DWAF, for ultimate consideration and gazetting by the Minister.

The WRCS also has bearing on a range of broader processes, given the socio-economic, political and ecological implications of the class. Accordingly, co-operation with all three spheres of government, participation of stakeholders and engagement with civil society, is required to ensure the appropriateness and acceptability of the WRCS and, ultimately, of the proposed class for a particular water resource. This implies that the classification process and the implementation of the class are founded on consultation, participation and co-operative governance to ensure socio-economic balance and sustainability, in addition to the technical elements of ecological sustainability.

To address the integrated nature of water resources utilisation and management, classification needs to take place at a catchment scale. This allows for the establishment of the relationships between various ecological resource units, economic utilisation and social well-being that are often exercised at different scales. The transfer of water between catchments adds complexity to the process of classifying water resources. Guidelines relating to the representation of designated classes with a catchment, region or the country, still need to be set.

6.11 Resource Quality Objectives

In recognition of the fact that water resources consist of more than just water and a channel in which the water flows, Resource Quality Objectives (RQOs) are clearly-defined goals relating to all aspects of water quantity, water quality and aquatic ecosystem quality, including the quality of in-stream and riparian habitats and aquatic biota. RQOs are expressed as numeric or descriptive (narrative) goals for resource quality within which a water resource must be managed. Similar to preliminary determinations of the Reserve and Management Class, preliminary determinations of RQOs are undertaken until the water resource has been classified and the class gazetted.

Resource water quality objectives (RWQOs) are considered a subset of RQO's, and can be defined as spatially and temporally incremental in-stream (or in-aquifer) water quality targets that:

- Give effect to the water quality component of RQOs, including ecological water quality requirements.
- May equal the RQOs but are more often set at a finer resolution (spatial or temporal).
- Dictate the level of impact and utilisation that may be collectively produced by upstream users.

Interestingly, the ecological water quality Reserve requirements are often less stringent than the RWQOs required by various water users. This emphasises the need to manage water resources in such a way as to ensure long term sustainable utilisation for all users and potential, future users. The RQOs provide an important element of catchment management plans, providing water resource managers and users with the overall objectives for ecological requirements.

The Management Class (MC), Reserve and RQOs, once available, should guide water resource management and be translated into actions in catchment management plans. The implementation and enforcement of the MC, ecological Reserve requirements and RQOs, are the next substantial challenge to ensuring the long-term sustainable use of water.

6.12 Resource Quality Monitoring

The implementation of RDM procedures can also only be effective if long-term monitoring and management programmes are in place. Currently, 11 resource-quality monitoring programmes exist in DWAF, aimed, primarily, at providing information on the status and trends of water resources (DWAF 2004d). The programme, specifically designed to provide information regarding the overall ecological status of water resources, is the National Aquatic Ecosystem Biomonitoring Programme. This programme consists of a number of sub-programmes for each water resource type. Currently, only the River Health Programme (RHP) component has been developed and is implemented.

The RHP primarily makes use of in-stream and riparian biological communities and habitats to monitor the response of the aquatic environment to human disturbances. As the RHP was specifically designed to reflect on the status and trends of aquatic ecosystem health, it is not specifically aimed at monitoring RDM compliance. It does, however, provide a useful background database for Reserve assessments and provides an indication of whether a resource class is being met. The RHP is also important in raising general awareness amongst the public around river health issues and the need to protect water resources.

Between 1998 and 2006, 14 river systems were assessed and reported on in State of River reports. Currently, nearly 200 rivers are being monitored. Analysis indicates that, for those rivers monitored, less than 2% are in a natural state, approximately 30% are in a good condition, 40% are in a fair condition and 30% are in a poor condition. However, monitoring compliance with Reserve and RQO requirements requires a more integrated and detailed approach to water resource monitoring than the RHP can provide.

6.12.1 Source Directed Controls and Water Use Authorisations

Source Directed Controls (SDC), such as water-use licencing, are measures primarily designed to control water-use activities at the source of impact. In practice, SDCs are the main tool to give effect to RDM. The MC, Reserve and RQOs must, therefore, guide the use of SDCs in both a proactive and reactive manner:

- Proactive management focuses on measures that will prevent or minimise future water quality or quantity problems, and
- Reactive management focuses on managing existing water resource problems to achieve the MC, Reserve and RQOs

In order for a decision to be made regarding the discharge of wastewater, the receiving water resource and the 'allocatable water quality' need to considered. The receiving water resource needs to remain "fit for use" to satisfy the most demanding water quality requirements of users of the resource.

In catchments with no water quality stress, a precautionary approach is applied by enforcing, particularly in respect of point discharges, uniform national minimum requirements. These may be relaxed in special circumstances, provided that the RQOs attached to a desired management class are maintained. In catchments with water quality stress, it is policy to (a) consider stricter requirements and/or (b) strictly regulate or prohibit unsustainable practices, to comply with RQOs and to achieve the desired management class.

Compliance or 'end-of-pipe' monitoring for both the wastewater discharge as well as in affected resources, should be included in water-use authorisations, when appropriate. These need to be closely-aligned with water quality RQOs and source management objectives. Such monitoring provides an important information base for subsequent well-focused, corrective actions in cases of non-compliance. The integration between the RDM and SDC strategies is, however, not being adequately achieved due to lack of synchronisation of decisions and lack of capacity to implement these provisions, with a resultant deterioration in the ecological state of many water resources.

6.12.2 Groundwater Source and Aquifer Protection Zoning Policy Development

Due to South Africa's widespread and highly-localised groundwater occurrence and use, it is not possible to protect all groundwater resources to the same degree. For this reason, a differentiated protection approach has been taken, based on the vulnerability and regional/local importance of aquifers. Aquifers are assigned management classes based on the potential yield, as well as the level of dependence that communities have on that water source. Aquifers that are a sole source of water for communities enjoy the highest level of protection.

A differentiated approach is also taken to control the disposal of water and wastewater, where land needs to be zoned according to its suitability for waste disposal. No waste disposal activities are allowed near the recharge zones of major aquifers or sole-source aquifers. The monitoring of the performance of waste disposal facilities and their associated pollution prevention measures is mandatory (DWAF 2000a).

6.13 Conclusion

"The measure of how successful we are at implementing policy and legislation must be the extent to which it successfully achieves its objectives" (Muller 2006). In this instance, success is measured by the extent to which the state of the environment improved – or deteriorated, while still providing for equitable and efficient water use and economic development. To undertake such an assessment requires monitoring and auditing as well as an adaptive management approach.

The ultimate success of the new water resource protection policy and legislation is likely to rest, not so much on economic and technical issues, but more on the capacity to develop and implement the systems. The necessary tools for the management and protection of water resources have been provided in the NWA, the National Water Resources Strategy, and in supporting policies and strategies. The methodologies to implement those tools are also progressively being developed, although they require simplification and streamlining. It is the human resource capacity that will be a critical limiting factor in achieving successful policy implementation. This is particularly true in fields of specialization such as water resource protection, where the necessary skills and experience are scarce, not only within government but throughout the country.

Water resource protection has experienced varying degrees of success around the country. Pockets of success have been achieved where a critical mass of technically-capable people and necessary resources are available. The need for water resource protection capacity building initiatives, to address capacity constraints is critical to ensure that the policy implementation process is less vulnerable.

The integration of resource protection initiatives is also a key challenge. As previously mentioned, in South Africa, DWAF is the government department with the line function responsibility for water resource management, while DEAT has the overall line function responsibility for biodiversity protection. Neither of these departments can effectively protect water resources, and their associated ecosystems, independently. Adequate water resource protection and the conservation of critically important aquatic biodiversity will only be achieved through a comprehensive analysis and understanding of the areas of overlap and the effective sharing of expertise.

Water resource protection will also not be successful unless it is integrated with other water resource management activities and decision-making processes. Water resource managers often see RDM as an activity to be undertaken in isolation, where the EWR are seen as providing information for assessing water availability or water quality problems within a water resource. The problems are not resolved through addressing the EWRs but, rather, through applying a number of other water resource protection and management measures. To optimise resource protection efforts, the relationships between the various bodies, dealing with water resource protection measures, need to be clearly-defined into a practical framework. Relationships between RDM and biodiversity conservation planning, EIAs and other protection initiatives outside of DWAF, also need to be clearly-defined.

The long-term success of implementing the national water legislation and policy will depend on government having the necessary levels of experience and expertise to guide this process and promote the integration of all resource protection initiatives. The tools, as provided in policy and legislation, and many of the methodologies required to implement RDM, as well as the other protection initiatives, are more than adequate. Unless the capacity-building issue is addressed as a critical national priority in the water sector, chances of long-term success in implementation will be limited. There also needs to be, with all levels of government, a willingness to work together on issues of water resource protection that cut across departmental boundaries and mandates.

References

Adams JB, Bate GC, Huizinga P, Taljaard S, Turpie J, Van Niekerk L, Whitfield AK,Wooldridge TH (2002) Development of a reserve methodology for estuaries: The Nahoon Estuary as a case study. Paper presented at the 4th Ecohydraulics & Enviro Flows Conference, Cape Town

Brown CA (1996, June) The need for instream flow assessments and how to meet this need. Report of a meeting held at the Water Research Commission, Pretoria

Brown CA, Joubert A (2003)Using multicriteria analysis to develop environmental flow scenarios for rivers targeted for water resource management. Water SA, Journal of the Water Research Commission (WRC) in South Africa

Brown C, Pemberton C, Birkhead A, Bok A, Boucher C, Dollar E, Harding W, Kamish W, King J, Paxton B, Ractliffe S (2006) In support of water-resources planning – highlighting key management issues using drift: a case study. WaterSA. Water Research Commission, Pretoria

Davies B, Day J (1998) Vanishing waters. University of Cape Town Press, Cape Town

De Coning C, Sherwill T (2004) An assessment of the water policy process in South Africa (1994 to 2003). Water research Commission Report TT232/04. Water Research Commission, Pretoria

DWAF (Department of Water Affairs & Forestry) (1996) South African water quality guidelines. Volume 7: Aquatic ecosystems. Pretoria, South Africa

DWAF (Department of Water Affairs & Forestry) (1999a) Resource directed measures for protection of water resources. Volume 4: Wetland ecosystems Version 1.0. Pretoria

DWAF (Department of Water Affairs & Forestry) (1999b) Resource directed measures for protection of water resources. Volume 5: Estuaries ecosystems version 1.0. Pretoria

DWAF (Department of Water Affairs & Forestry) (1999c) Resource directed measures for protection of water resources. Volume 6: Groundwater component version 1.0. Pretoria

DWAF (Department of Water Affairs & Forestry) (2000) Policy and strategy for groundwater quality management in South Africa. Pretoria

DWAF (Department of Water Affairs & Forestry) (2001a). Draft: Methods for assessing water quality in ecological Reserve determinations for rivers. Version2. Draft 13. Pretoria. South Africa

DWAF (Department of Water Affairs & Forestry) (2002) Water resource protection policy implementation, methodology for the determination of the preliminary ecological reserve for estuaries, Version 2. Pretoria, South Africa

DWAF (Department of Water Affairs & Forestry) (2003) RDM module 1: Introductory Module. Pretoria

DWAF (Department of Water Affairs & Forestry) (2004). A 5-year Water Resources Quality Monitoring Plan. Department of Water Affairs and Forestry Internal Report. ISBN No. 0-621-35488-0. Pretoria

DWAF (Department of Water Affairs and Forestry) (2004a). National Water Resource Strategy. First Edition. Pretoria. South Africa

DWAF (Department of Water Affairs & Forestry) (2006) A toolkit for water allocation reform. A manual to achieve race and gender equity in water allocations. Pretoria

Hughes DA, Münster F (2000) Hydrological information and techniques to support the determination of the water quantity component of the ecological reserve for rivers. Report to the Water Research Commission by the Institute for Water Research, Water Research Commission – South Africa. WRC Report No. 867/3/2000

Louw MD, Huggins G, Birkhead AL, Long GC, Van Wyk NJ (2002) Environmental flows and the reserve: theory into practice. Presentation at the 4th ecohydraulics and enviro flows conference, Cape Town

Minister of Water Affairs and Forestry (2006). Benefit Sharing in Transboundary Waters – The SADC Approach and the South African Experience. Speech by Mrs L. Hendricks, Min of Water Affairs and Forestry, Stockholm Water Symposium, Stockholm, Sweden

Muller M (2006). Transforming water law to achieve South Africa's development vision: A case study in national law

Palmer CG, Rossouw N, Scherman P-A, Muller WJ (2002) The development of methods to integrate water quality into riverine environmental flow assessments. Paper presented at the 4th ecohydraulics and enviro flows conference, Cape Town

RSA (2003) Strategic Framework for Water Services. Government Printer, Pretoria, South Africa

Tharme RE (2002) A global perspective on environmental flow assessment: Emerging trends in the development and application of environmental flow methodologies for rivers. Paper presented at the 4th ecohydraulics and enviro flows conference, Cape Town

World Wildlife Fund (WWF) (2003). Towards a just South Africa: The Political Economy of Natural Resource Wealth (Ed. D. Reed and M. de Wit). WWF Macroeconomics Program Office, USA

Chapter 7
Catchment Management Agencies: A Case Study of Institutional Reform in South Africa

Eiman Karar, Gugu Mazibuko, Thomas Gyedu-Ababio, and Derek Weston

Abstract This chapter highlights the decentralisation of water resources management to a local level through the establishment of Catchment Management Agencies (CMA). The first 8 of 19 proposed CMAs have been established and a number of institutional challenges have been identified. Stakeholder participation in the establishment of CMAs has been largely successful, but the process of establishment has been too slow and further delayed by a review of the number of CMAs to be established. The Inkomati experience shows that a lack of coherent support from DWAF and building the credibility of the institution in the water management area have been major challenges. The fact that DWAF is the driver and initiator of CMA establishment placed huge strain on the already-limited human capacity that had to perform these tasks in addition to daily water resource management activities.

Although the decentralisation-based reforms for water management define a critical role for communities and users at large, the state must, and will continue to, play a fundamental role because of its responsibility for managing water as a public good and for ensuring redress, equitable allocation of water and equitable representation in decision-making. As a water-scarce country, the need to safeguard the sustainable provision of ecosystem goods and services and to protect the interests and welfare of all users, especially the poor, women and the disabled becomes critical.

E. Karar (✉)
Water Research Commission, Pretoria, South Africa
e-mail: eimank@wrc.co.za

G. Mazibuko
Pegasys Strategy and Development, Pretoria, South Africa
e-mail: gugu@pegasys.co.za

T. Gyedu-Ababio
South African National Parks, Pretoria, South Africa
e-mail: thomasga@sanparks.co.za

D. Weston
Pegasys Strategy and Development, Pretoria, South Africa
e-mail: Derek@pegasys.co.za

B. Schreiner and R. Hassan (eds.), *Transforming Water Management in South Africa: Designing and Implementing a New Policy Framework*, Global Issues in Water Policy 2, DOI 10.1007/978-90-481-9367-7_7, © Springer Science+Business Media B.V. 2011

Keywords Decentralisation • Governance • Public participation • Catchment management agencies • Institutional change • Capacity building

7.1 Introduction

As has been discussed in Chapter 1, governance of the water sector in South Africa has been strongly influenced by political and socio-economic changes. During the period after the first democratic election in 1994 considerable effort was applied to redress the lengthy history of conquest, expansion and exclusion of people from access to water and importantly, in the governance context, the exclusion of people from the actual management of the resource. This resulted in the promulgation of the National Water Act (Act No 36 of 1998).

The previous Water Act of 1956 favoured a dominant white minority that had privileged access to land, water and economic power. Under the pressures of settler expansion and economic development, water was captured as a resource for the white minority. This resulted in the post-1994 democratic government being confronted with a situation in which the majority of South Africans had been denied access to water for productive use or access to the benefits from the use of the nation's water. The NWA seeks, *inter alia,* to address these past imbalances.

The importance of equitable access to water has formed part of the political debate since the development of the Freedom Charter and the establishment of the new South African Constitution (De Coning and Sherwill 2004). As part of the process of addressing the inequity in access to water and the exclusion of people from the decision-making process around water, the White Paper on a National Water Policy for South Africa (RSA 1997) sets a framework for institutional reform in the water sector: - "The reforms must reflect the requirements of fairness and equity, values which are cornerstones for South Africa's new Constitution. It must also reflect the limits to the water resources available to us as a nation". The vision for water institutional reform is to create institutions that will ensure the sustainable and equitable allocation of water and better water management that will contribute to a "better life for all".

The National Water Act enables a new set of institutional arrangements, including the establishment of Catchment Management Agencies to perform delegated water resource management functions, which are a move away from the historically centralised management of water resources. A key intention behind the establishment of CMAs was that, through decentralised resource management, stakeholders, especially those previously excluded from access to water and to the benefits derived from water, should be able to participate in water resource management decision-making.

It was recognised that water, especially in the water-scarce Southern African context, would play a central role in the country's growth and development. Thus, section 2 of the NWA states that the purpose of the Act is "to ensure that the nation's water resources are protected, used, developed, conserved, managed and controlled in ways that take into account:

- Meeting basic human needs of present and future generations.
- Promoting equitable access to water.
- Redressing the results of past racial and gender discrimination.
- Promoting the efficient, sustainable and beneficial use of water in the public interest.
- Facilitating economic and social development.
- Providing for growing demand for water use.
- Protecting aquatic and associated ecosystems and their biodiversity.
- Reducing and preventing pollution and degradation of water resources.
- Meeting international obligations.
- Promoting dam safety.
- Managing droughts and floods.

And, "for achieving this purpose, to establish suitable institutions and to ensure that they have appropriate community, racial and gender representation."

7.2 Water Resources Management Institutional Arrangements

As for any country that requires new sets of institutional arrangements for water management, the challenges are numerous. For many countries, the institutional inertia is significant enough to effectively prevent any chance of reform. Countries in the European Union are struggling with the challenge of bringing in more participative and decentralised institutions. However, in the South African context, the timing of the societal changes, together with the policy and legislative reform, created an opportunity to introduce innovative approaches.

The institutional model for water management in South Africa is reflected in Fig. 7.1. As can be seen from this figure, the model provides a complex arrangement with a considerable number of institutions with varying relationships, some of which focus on water resources management and some on water services. The Department of Water Affairs and Forestry remains the core that is ultimately responsible for ensuring the delivery of effective and sustainable water services, and for ensuring the equitable, sustainable and efficient use of water resources.

This chapter focuses, specifically on the experience relating to the establishment of Catchment Management Agencies (CMAs), and on some of the lessons that have been learned in the process to date. It is important to note, that while the Act was passed in 1998, by the end of 2008 only eight CMAs had been established, and only two can be considered functional. It is, therefore, still early in this process and the establishment process still has a long way to go.

As the custodian of the country's water resources, the Department of Water Affairs and Forestry remains the ultimate accountable entity for the sustainable, equitable and efficient use of water resources. However, the Department is not required to perform all water resource management functions itself, and is mandated, under the Act, to delegate functions to CMAs and other water management

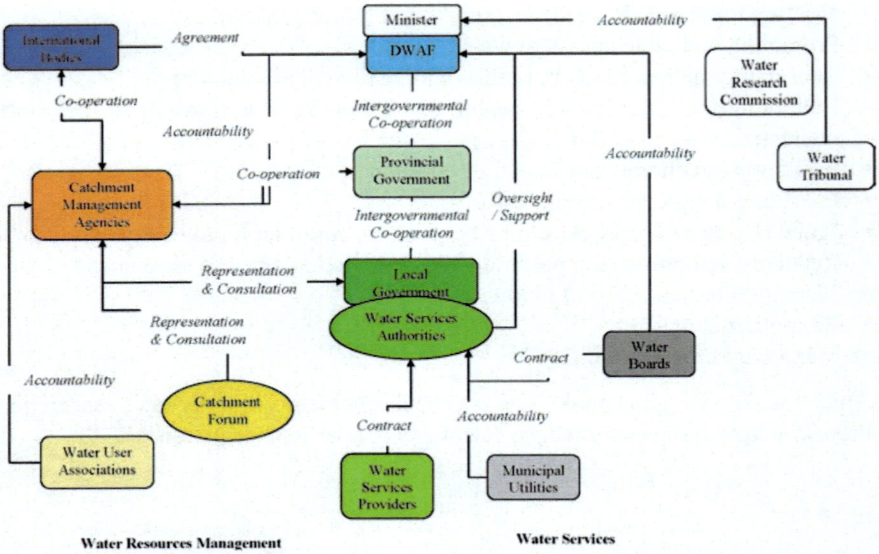

Fig. 7.1 Institutional arrangements for water management in South African

institutions, such as Water User Associations, as and when appropriate. The delegation of functions to CMAs is based on the principle of subsidiarity, which requires the management of resources at the lowest appropriate level, as described in Agenda 21. The Department, however, remains responsible for ensuring that these water management institutions deliver according to national policy and legislation. The Department must, therefore, take up a stronger regulatory and oversight role as functions are delegated to water management institutions.

The purpose of Catchment Management Agencies is described in the introductory notes to Chapter Seven of the National Water Act: "The purpose of establishing these agencies is to delegate water resource management to the regional or catchment level and to involve local communities, within the framework of the national water resource strategy, established in terms of Chapter 2. Whilst the ultimate aim is to establish catchment management agencies for all water management areas, the Minister acts as the catchment management agency where one has not been established. Where the necessary capacity does not exist to establish a catchment management agency, an advisory committee may be appointed under Chapter 9 to develop the necessary capacity as a first step towards establishing an agency."(RSA 1998)

Shortly after the promulgation of the National Water Act, and after a process of public consultation, 19 Water Management Areas were declared, their boundaries having been determined according to the Act, which requires the Minister, in determining a water management area, to take into account watercourse catchment boundaries, social and economic development patterns, efficiency considerations, and communal interests within the area in question (RSA 1998) (see Fig. 7.2).

During this early process, various institutional challenges started to be recognised, including achieving a balance between the size of the water management area

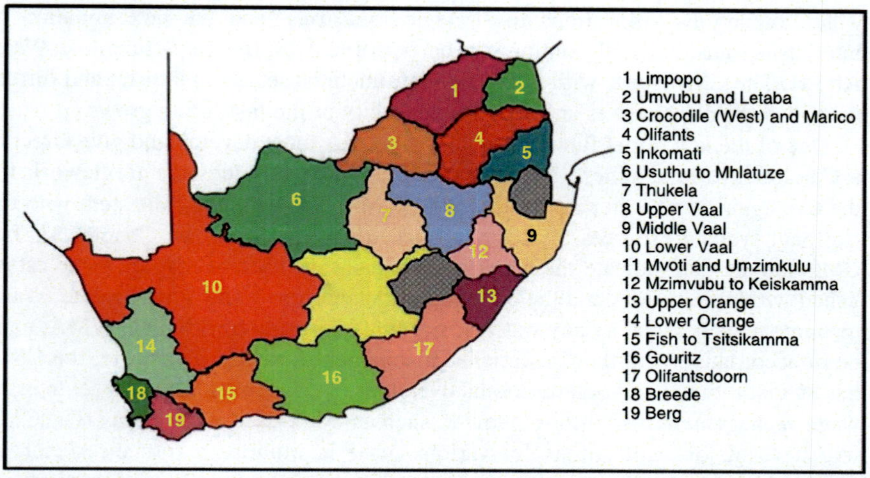

Fig. 7.2 The 19 Water Management Areas declared under the National Water Act (Source DWAF)

and having a local institutional presence, the disjuncture between hydrological and political boundaries, the mismatch of surface water boundaries and those of groundwater aquifers, and the relationships between water management areas and the larger basins of which they formed part.

By the end of 2009, eight CMAs had been formally established through the Government Gazette, but only two Governing Boards and two CEOs had been appointed, and whilst the Breede Overberg was currently appointing its first line managers, only one, the Inkomati CMA, had a significant staff compliment in place under the CEO.

Each regional office of the Department of Water Affairs and Forestry put in place a process towards the establishment of CMAs, although these processes varied considerably. Some took a very direct route and this is reflected in the results above. Others placed more emphasis on establishing forums and water-user associations as a platform to support the CMA establishment process at a later stage.

While it was originally agreed that a CMA would be established in each WMA, a current institutional review process has proposed that no more than nine CMAs should be established, with the 19 WMAs clustered under them. The establishment of further CMAs has been put on hold pending the finalisation of the current review. Certainly, the reduced number of CMAs is intended to resolve capacity issues in staffing the CMAs and in oversight and regulation of the CMAs. Equally, and quite importantly, the reduced number of CMAs would provide for stronger governance within the new institutions, which is currently a concern within the Departments of Water Affairs, Public Service Administration and National Treasury.

Once a CMA is fully functional, it should be responsible for most water resource management functions within its area of jurisdiction, including the authorisation of water use. Certain strategic functions, such as the determination of international water allocations, allocations to strategic industries, and the determination of interbasin transfers, will continue to be performed by the Department of Water Affairs. The CMA starts with a set of initial functions set out in the Act and further functions are delegated as and when the capacity of the institution grows.

One of the key initial functions is that the CMA must develop and give effect to a Catchment Management Strategy (CMS), which provides the framework for the management of water resources in the WMA, and which is consistent with the National Water Resources Strategy (NWRS) (DWAF 2005) (see Chapter 8). The CMS should also indicate the water management institutions that are to be established within the WMA and the participatory environment that it will create. Water resource management, by any water management institution within the WMA, must be in accordance with the Catchment Management Strategy. Therefore, the CMA has an institutional and organisational oversight responsibility with respect to other water management institutions below it, such as water-user associations. The CMS will have major implications for various other institutions across the sector, as reflected in Fig. 7.1. These institutions, therefore, also become critical participants in the CMS development process.

The CMA has a number of functions and responsibilities that require it to interact with various organisations and stakeholders at national, regional and local level to achieve the equitable and sustainable management of water resources. The interactions can be defined in the following manner:

Oversight and Accountability Interaction: the CMA Board, as the Accounting Authority, must report annually to the Minister of DWAF, as the Executive Authority, on its activities. Furthermore, the CMA must develop a business plan that should state the objectives for the next financial year and the financial requirements to achieve the objectives in the business plan.

Water Use Management: one of the key functions of the CMA is to ensure the sustainable, efficient and equitable management of water within the WMA. Water users, who are authorised to use water through relevant authorisation tools, must comply with the conditions of authorisation. The CMA must ensure that these conditions are complied with, including by other organs of state such as local government, and private sector water users such as industry, mining, and the irrigation sector.

Integrated Planning: each CMA must prepare a Catchment Management Strategy (CMS), which provides the framework for the planning and management of water resources within its area of jurisdiction. The Act requires that the strategy take into account all other development plans that impact on the management of water, including those of local government (through the Integrated Development Plans) and provincial government (through the Provincial Growth and Development Strategies) (cf. Chapter 7).

Arrangements for Stakeholder Involvement: one of the initial functions of the CMA is to co-ordinate the activities of stakeholders within its area of jurisdiction and to develop institutional arrangement for the participation and empowerment of stakeholders in the management of water resources.

Each CMA needs to ensure the existence of the institutional arrangements for the participation of stakeholders. A range of structures used to promote participation of stakeholders is briefly discussed below:

7.2.1 Catchment Management Committees (CMC)

The NWA provides for the formal establishment of committees by the CMA, either to advise it, or to perform any of its functions within a specified area. Catchment Management Committees (CMCs) must have representation from the CMA Governing Board and/or employees of the CMA, but may include other specialists and/or stakeholders. The CMC is established under the Act, and is different from a sub-committee of the Governing Board.

A geographically-based CMC would focus on key issues within a specified sub-catchment area of the WMA. Area-based CMCs provide a mechanism for communication, co-operation and decision-making between stakeholders and the CMA. Issue-based CMCs could consider specific issues for the entire water management area, such as water quality, river health or specific water use activities, for example. To date, however, no CMCs have been established.

7.2.2 Catchment Management Forums (CMF)

Catchment Management Forums, which are not specified in the Act, nonetheless, provide opportunities to facilitate stakeholder participation in a number of activities, including the establishment of the CMA, the development of the catchment vision, the development of the Catchment Management Strategy, and the determination of Resource Quality Objectives (cf. Chapter 5).

The Catchment Forum is, in essence, a structure in which stakeholders can choose to participate on a voluntary basis, and through which, the CMA can consult with and inform Forum members of key actions. The CMA can ask the Forum to perform some functions on its behalf (especially in situations where a CMC is not in place or is not functional). A number of Forums have been established across the country, and many have been extremely active in water resources management functions in their areas.

7.2.3 Water User Associations

Water User Associations (WUAs) are co-operative associations of water users established under the National Water Act to undertake water-related activities for the mutual benefit of their members within a designated area. WUAs are mainly, but not only, established to manage local water infrastructure, e.g. irrigation water supply schemes and to implement management decisions agreed upon between the members. Importantly, the WUA has a strong role to play in water-use charge collection and as such plays a pivotal role in the longer term viability of the CMA.

The mandate of a WUA may also be stretched to perform other water resources management and water services functions at local level.

7.3 Establishment of CMAs

According to the National Water Act, the establishment of a CMA can be initiated by the Minister or any interested citizen. The individual or group would have to demonstrate how all users and affected and interested parties were consulted and involved in the process. According to the guidelines for proposal submission drawn up by DWAF, a proposal needs to be compiled and handed to the Regional Office who, in turn, would formally submit it to the National Office. To date, all CMA establishment proposals have been driven by the Department and not by stakeholders, recognising that not all stakeholders have the required resources available to take up such a process.

In all CMA processes to date a reference group was established to support the development of the proposal and work with the details of preparing that document. The reference group that writes the proposal is derived from the arrangement as illustrated in Fig. 7.3.

In accordance with the current guidelines, the proposal must include, among other information, details of the boundaries within which the proposed agency will operate, the water resources and existing infrastructure in the area, and the ways in which they are managed. Most essential is that the proposal reflect the vision of the stakeholders as to how they see the CMA functioning.

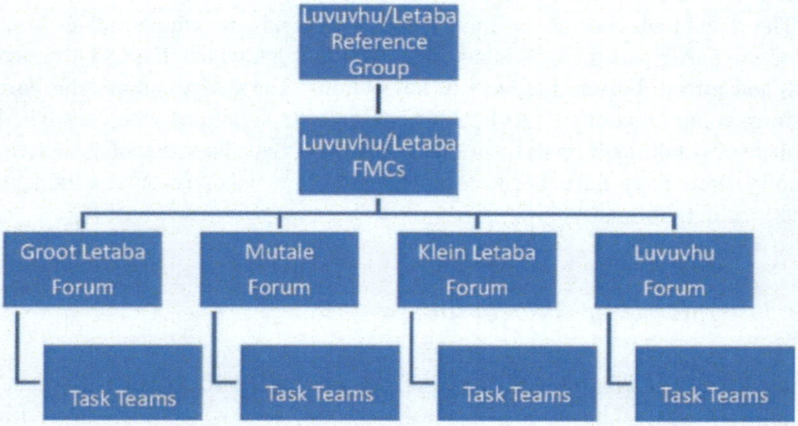

Fig. 7.3 Example of structure of CMA establishment reference group from Luvuvhu/Letaba water management area

Public involvement contributes to establishing the legitimacy of the institution, assists the advisory committee in understanding the critical issues around constructing an appropriate Governing Board, and builds a foundation for the CMA to promote and continue public involvement in water resource management. Accordingly, the extent to which stakeholders have been involved in the development of a proposal to establish a CMA is one of the most important criteria against which the Minister judges the merit of the proposal.

Stakeholder participation can be initiated by distributing information to create awareness. The establishment of representative forums helps to develop constructive and trusting relationships between water resource managers and the public, with the aim of forming a common vision and understanding of the future agency's role and functions. The assessment of the role and function leads to an assessment of the level of staffing required to perform these functions, and from which the agency's operational costs can be derived. Comparing these costs with the anticipated revenue from water resource management charges, determined in accordance with the pricing strategy (see Chapter 8), can determine the likelihood of the CMA being financially self-sufficient or the level of subsidy required.

The proposal is then reviewed and evaluated against specific criteria and a recommendation made to the Minister on the outcome. If successful, the Minister places an announcement in the Government Gazette that the proposal is available in accessible locations for comment, for a minimum period of 60 days. However, experience has shown that this is not sufficient and proposals are now being gazetted for 90 days. In most instances, the translation of a summary proposal has proven essential to allow for non-English speaking users to comment as well. The Department's Regional Offices have also held work-sessions to provide a platform for comments. This has proven most helpful for more marginalised communities.

Most stakeholders seem to view the comment process as an opportunity to air their discontent with aspects of the proposal and process, rather than submitting positive comments.

After the comment period, if the Minister is satisfied that due process has been followed, he/she announces the establishment of the CMA in the Government Gazette. DWAF then establishes an external Advisory Committee (AC) to advise the Minister on the possible composition of the Board and the identification of representative institutions to nominate members to the Board. Membership of the Board is governed by section 81 of the Act, which states that the Minister, in appointing the Board, must aim to achieve a balance among the interests of water users, potential water users, local and provincial government and environmental interest groups, and must consider the extent to which local government should be represented on the board (RSA 1998).

The Advisory Committee members are selected both from the WMA and from people outside the area who have appropriate expertise and interest. The role of the Advisory Committee is to identify major interest groups in the WMA and to call for public meetings where possible Board composition is tested. Based on this work, the Advisory Committee makes a recommendation to the Minister on the composition of the Board.

The recommendations on the sectors that need to form part of the board prompt the Minister to communicate with those sectors to nominate three potential representatives from whom she can choose one, in order to attain a gender, racial and technical balance in the final selection.

Ministerial approval of the proposal paves the way for the appointment of the Governing Board. Once the Governing Board has been appointed, they embark on a process of institutional development. During this initial phase, the new CMA appoints a Chief Executive Officer, develops its first business plan and strategic vision, and starts the process of developing institutional policies and structures. With this in mind, the establishment and full empowerment of catchment management agencies, in all water management areas, will take some time to achieve, even with a reduced number of CMAs. In the meantime, the Department will continue to manage the water management areas.

As custodian of the nation's water resources, it is incumbent upon DWAF to provide the necessary support to the new CMAs whilst continuing to manage the resource. This means a number of things:

- The Board members, representing different sectors, do not necessarily have much expertise in water resource management, so the Department must ensure that their capacity is built in this regard.
- The institutional development process needs to be supported financially by government funds and not by users.
- Technical and governance support needs to be provided and is intended to last, at least, until the Board has articulated its first business plan and has appointed its Chief Executive Officer and, probably, the first line of senior managers. From then onwards, the state assumes its regulatory role.
- The Department needs to actively support the CMS development process.

During the transition between the establishment of CMAs and full functionality, the Department and the CMAs have to work closely together. The respective roles will change as powers and duties are delegated and assigned to the CMAs, and it is essential for roles and functions to be clearly-defined at each stage of the transition so as to avoid conflict. Eventually, the Department will be responsible only for oversight, regulation and general support of the agencies. In the Inkomati CMA it has taken some time for strong working relationships to be developed between the CMA and the DWAF Regional Office, with the associated clarity on roles and responsibilities.

The type of CMA to be established is determined by a number of factors ranging from the challenges faced by the WMA, to the priorities and capacity within the WMA. Hence, the CMA can be a rural, agricultural, community-dominated, small CMA, such as in the Olifants Doorne WMA; a rural, agricultural, industrial, mining, medium-sized CMA such as in the Inkomati and the Olifants WMAs; or an urban-dominated, large CMA such as in the Crocodile-Marico and Mvoti to Mzimkhulu WMAs. Bearing in mind these nuances, and the local context, a 'one size fits all' approach is not appropriate and this has resulted in varying degrees of progress regarding CMA establishment and development.

The CMA Governing Board, using a Board Charter agreed to with the Minister, is expected to provide strategic direction to the CMA. The Board Charter should, to a certain extent, provide the key strategic objectives, that the Board should achieve, on behalf of the Minister and the Department. Based on the understanding of these strategic objectives, the Board should also design and develop the CMA in a manner that will achieve the development of the objectives. Furthermore, whilst the NWA does not stipulate the requirement, it is becoming clear that a formal Service Level Agreement between the CMA and the Department will prove essential.

7.4 Oversight and Regulation

With the decentralisation of water resource management comes the need for strong oversight and regulation of the decentralised institutions implementing the legislation. This requires clear policy direction, and clear roles and responsibilities for the water management institutions. These instructions are, arguably, articulated in the National Water Act of 1998 (Asmal 1998). According to the defined rules, the Accountable Authority has the liberty to delegate some of its responsibilities but not the authority to delegate itself. For the Department to delegate functions, it must define clear expectations and how performance will be measured. This also requires of the Department, and indeed the water sector, to understand what it really means to oversee and to regulate.

Oversight, in the South African context, is defined in terms of monitoring/supervision, support and regulation. Support must be provided to the new water management institutions until they are functional authorities that can be regulated. they are responsible for decisions and actions against clearly-prescribed performance requirements. This will happen in a gradual and progressive manner. As such, initially, there is joint responsibility allowing for the lines of accountability to become somewhat blurred.

This duality in the roles played by the state can bring about difficulty in the clear definition of accountability and in the sourcing of the required regulatory skills and resources during the transitional period. The current, apparent trend is that the same resources deployed in establishing and supporting CMAs are responsible for regulating them. This is not surprising when realising the severe shortage of skills to populate both roles. In many instances, the same people play both roles and this creates a distinct challenge.

As a result, the appropriate institutional arrangements for regulation are under discussion, including whether an independent regulator is needed. The main motivation for this is the separation of the support role from regulation and enforcement roles, and to ensure that proper human resources are deployed to perform the regulatory functions on the state.

CMAs will require oversight and regulation in terms of how they perform against key functionality measures and how they deliver the services for which they are responsible. Furthermore, and importantly, there needs to be oversight and regulation around issues of governance. This would include corporate governance issues as well as those of organisational governance.

Corporate governance of water management institutions, such as CMAs, is based on the guidelines and regulations set out for public entities by the Departments of Public Service Administration and National Treasury.

Regulation of social equity and technical regulation are probably the most difficult aspect of the newly-assumed roles. Central to the NWA are social equity and redress, which cut across all functions. The skills required for social and technical regulation are not readily available. Efforts have to be invested in grooming and cultivating such expertise in partnerships with tertiary institutions and by learning from other countries.

CMAs, at a sub-regional scale, will be responsible for regulating water use and local water management institutions, such as water user associations. Hence, oversight happens at different levels and the dichotomy between support and regulation will appear at all levels. Separating these roles is essential whether within the same organisation or through establishing separate organisations.

Within this regulatory role, the state has a particular responsibility to ensure that water resources management results in both sustainable and equitable water use that redresses the inequities created by apartheid. As the custodian and protector of the poor and the marginalised, the state is in the best position to enable silent voices to be heard. The proper level of governance will need to be interrogated and the water sector faces some significant challenges in ensuring that this takes place. A clear example of the nature of such challenges includes the relationship between CMAs and local government. Certainly, there is more and more need for partnership and close collaboration between CMAs and local government in enforcing compliance with water quality and resource protection measures as well as social equity deliverables.

7.5 Key Lessons

The following section draws out some of the key lessons in the establishment of effective CMAs, based on experiences from the establishment processes to date.

7.5.1 Wall-To-Wall CMAs?

While the National Water Act does not require the establishment of CMAs in all water management areas, this has been the approach adopted by the Department to date. Thus, rather than wait for stakeholders to express a need for a CMA, the Department has initiated and driven all of the processes for the establishment of CMAs. However, two factors make this approach questionable. The first is the shortage of capacity for staffing CMAs, and the second is the challenge of establishing CMAs in areas where they will not be financially viable.

Research done by the Department shows that not all CMAs will be independently financially viable. A variety of factors impact on the viability of CMAs, particularly in water management areas dominated by agriculture, and which show relatively low water use, such as the Gouritz and Olifants-Doorn water management areas. The continued capping of the water-use charges for irrigated agriculture has contributed to the lack of financial viability of some CMAs. Non-viable CMAs will require subsidies from the state on an ongoing basis.

However, in other areas where pollution and water-quality problems present a particular challenge, the introduction of the waste discharge charge system (see Chapter 9) will improve the financial viability of some CMAs. The Crocodile West Marico, Upper Vaal, Middle Vaal and Mvoti to Mzimkulu water management areas will obtain some 50% of their income from the waste discharge charge system. Delays in the introduction of this system will affect these CMAs in the short term.

There are also concerns regarding the ability to recruit sufficient staff for all of these CMAs due to the general shortage of experienced technical personnel in the country. This is particularly pertinent for CMAs in marginal and underdeveloped areas. Since one of the primary drivers behind the establishment of CMAs is enabling stakeholders to participate in decision-making, it must be asked whether this cannot be equally well achieved by putting in place a stakeholder-based Advisory Committee (as enabled by the Act) which can advise the regional office of DWAF in terms of the management of water resources.

This issue links closely to the issue of scale and number of CMAs as discussed below.

7.5.2 *An Issue of Scale*

This issue was initially discussed during the deliberations on the water management areas and their boundaries. During those discussions, some stakeholders made it clear that they expected to have a CMA on each major catchment. Whilst this would present challenges in terms of financing and staffing, importantly, it would also place a considerable oversight burden on the Department, simply due to the number of CMAs that it would entail. A balance was, therefore, sought between the number of institutions and the requisite oversight requirement, with the ability to provide a local presence and enable a span of control.

In the latest institutional review, these concerns were again raised and consensus appears to have been to cluster the water management areas under nine CMAs, rather than the original 19. Increasingly, the Department of Water Affairs and Forestry, and the National Treasury have been concerned about being able to ensure that there is solid governance within CMAs. Fewer Governing Boards would present a lower oversight burden and would link weaker water management areas with stronger and more capacitated water management areas so that there would be a stronger possibility of obtaining a more "professional" Board. The institutional review also provided the space for a re-think on some aspects of the Act, including the requirements on the membership of the Governing Boards.

7.5.3 Stakeholder Consultation

A critical element of the CMA process is stakeholder participation. What is important in the South African context of huge inequities in relation to class, race and gender, is not only the identification of stakeholders, but also the empowerment of marginalised stakeholders so that they can participate effectively in processes.

There are three particularly important categories of stakeholders that must be considered. The first category comprises water users in the water management area, who will ultimately be regulated by the CMA. This group of stakeholders includes municipalities and the irrigation, power, industrial and mining sectors, as well as the emerging and developmental water users who have very real and specific challenges. The second group of stakeholders is made up of other organs of state that serve as partners to the CMA in assisting it to manage water effectively and sustainably, such as Departments of Environmental Affairs, conservation bodies, and water infrastructure agencies. The third group includes civil society groups who are largely focused on protecting environmental or social rights, such as NGOs, CBOs and Faith-Based Organisations.

For a number of reasons, the CMA proposal development process has, on average, taken more than 12 months and there have often been long intervals between stakeholder meetings, resulting in challenges in maintaining stakeholder interest and continuity in stakeholder representation. In some instances, the establishment process has been hampered by stakeholder unavailability or new stakeholder representatives who need to be brought up to speed. In some areas, the use of formalised stakeholder bodies, such as Catchment Management Forums, has facilitated improved stakeholder consultation.

A particular challenge has been ensuring continuity amongst the representatives of marginalised communities. Often these representatives do not have the financial means to attend meetings, and the Department developed guidelines for public participation that specify mechanisms to be used to ensure that such groups are engaged with and are able to participate in consultative processes. The Department also developed a policy on the payment of travel and attendance allowances for representatives of poor communities to recompense them for the lost income arising from attending workshops and meetings.

Equally important, has been the process of empowering marginalised community representatives to participate alongside extremely well-resourced and powerful water users. In the KwaZulu-Natal, CMA processes were dealt with by having pre-meetings with marginalised community representatives. These meetings would serve as capacity building sessions prior to the main stakeholder meeting. Whilst this did have some impact, the two meetings were held on the same day, which provided little space for the "digestion" of information. The approach in the Western Cape was slightly different in that a special capacity building session was held a week or so in advance of the main stakeholder event. Whilst this provided time for the "digestion" of material, it also placed an extra burden of travel and time on these groups, many of whom could ill-afford either. In both instances, special attention was provided, through the effective facilitation of meetings, to ensure that these groups had a chance to have their say.

In the Inkomati, use of Catchment Management Committees and Catchment Management Forums has been chosen as the optimal method for stakeholder consultation. The Catchment Management Forums met to discuss water resource management issues and were, indeed, involved in the process to establish the Inkomati Catchment Management Agency. However, as with many forums around the country, the composition of these Forums is not always as fully-representative as one would wish. In the Inkomati, the major catchments clearly have different stakeholder and water-use patterns which require tailored and appropriate forum structures. The important thing to note is that, with the more localised presence of the CMA, the forum representivity and involvement is improving.

7.5.4 Board Composition

In the establishment of a CMA, an Advisory Committee (AC) is appointed to advise the Minister on the composition of the CMA Board. Experience from the CMA processes, to date, indicates that the membership of the Advisory Committee should reflect local knowledge of the water management area, an understanding of government planning and broad strategic objectives of government. The members reflecting local knowledge should, preferably, be from one of the reference groups that developed the proposal as they have better knowledge of the area on issues ranging from environmental challenges, water management challenges and requirements. Experience has shown that failure to reflect this knowledge may have a limiting factor on the reflection of a wide range of issues on the proposed CMA Board composition.

The Advisory Committee also has to workshop the proposed Board composition with stakeholders within the water management area. Experience has shown that special attention should be given to ensuring the inclusion of marginalised communities in this process. The process should be given sufficient time to ensure that all stakeholders are on board with regard to the unfolding process, what it means for them and how can they influence it to reflect their issues or concerns.

As has been mentioned, the National Water Act requires the Board of the CMA to be representative of water users, local, and provincial government, and environmental interest groups. However, there are concerns regarding this approach. Firstly, having to ensure representation of all key stakeholders and provincial and local government results in an unnecessarily large Board (e.g. the Board of the Inkomati CMA consists of 13 members). There is also a tension for Board members between acting in the interests of the CMA as is required by sound corporate governance principles, and representing the interests of their own constituencies. Furthermore, a stakeholder board does not necessarily bring to the table some of the key fiduciary, legal and technical skills required for the effective functioning of a Board. The Department has attempted to create an effective blend of skills and representation in the Boards, but the skills required are not necessarily those brought to the table by stakeholder representatives.

Finally, the Board plays a critical role in determining the strategic direction of the CMA, and in overseeing policy development and implementation. Due to the current, skewed water allocation landscape, it is expected that the well-capacitated, large users and interest groups will dominate the Board. While Advisory Committees have tried to find a balance between current and potential users, the large users tend to be more powerful because of greater skills, financial and technical resources. As large users, their financial contributions to the CMA, through water-use charges, are significant and it possible, as a result, that they can exert undue pressure in setting terms and making decisions. In order to manage this, strong regulation from the Department is required, but the capacity for such regulation is not yet in place and is unlikely to be able to engage with CMAs based on daily decision-making.

As a result, the Department is considering amending the Act to enable the appointment of smaller, more highly-skilled Boards that have limited stakeholder representation. The representation of stakeholders could be achieved through Catchment Management Committees or Advisory Committees that provide input and guidance both to the Board and to the management of the CMA.

7.5.5 Institutional Change and Capacity Building

Since CMAs are agencies performing water resource management functions on behalf of the Department, DWAF must provide the necessary institutional development support, especially in the initial stages of CMA development. As part of the initial institutional development for the CMA, there are specific activities that need to take place, such as the transfer of staff from DWAF to the CMA,[1] change management and marketing, development of systems, transfer of budgets, and so on. The Department must be prepared for these activities. The time and resources required for this task should not be underestimated.

Unfortunately, in the establishment of the Inkomati CMA, the Department was not fully prepared. The process had implications for a range of sections of DWAF, and required decisions to be made in the Corporate Services Branch, the Regional Office, and the Policy and Strategy Branch. Co-ordination of these processes was weak, resulting in the poor alignment of processes in terms of approach and timing. Processes for the transfer of staff and funds were particularly contentious. In the Inkomati CMA establishment, there was a failure to ensure that effective financial systems were in place for the CMA in order to facilitate the transfer of funds. This, combined with lack of clarity regarding the actual funds, due to the CMA, from the water-use charges in the WMA, resulted in significant delays in the transfer of funds. Lack of clear direction on the transfer of staff may well have contributed to resistance from certain DWAF staff to their potential transfer to the CMA.

[1]A Proto-CMA is a functional structure established in the DWAF Regional Office in preparation for transfer of staff and functions to the CMA once it is established.

The transfer of staff is a complex matter that involves negotiations with Trade Unions, particularly with regard to staff benefits and job descriptions. In the Inkomati CMA instance, a Memorandum of Agreement was signed between DWAF and the Trade Unions to assist this process. This agreement was signed in mid-2007 and the transfer of staff was due to take place in August 2008. At the time of writing, the transfer had still not taken place.

Progress with the establishment of subsequent CMAs has been smoother because of the lessons learned in the Inkomati process. Certainly, it helped in the Breede Overberg CMA process, although the nature of the water management area and the Regional Office resulted in a different process with a different flavour. The issue of staff transfers in this water management area were not as complex and the CMA had to appoint new staff, with very few staff being transferred from DWAF.

The other important issue in the transfer and secondment of staff (staff migration) is that DWAF needs to identify, well in advance, the functions that will be transferred to the CMA so that staff responsible for those functions in DWAF are transferred timeously. This means that functions currently residing within DWAF, and a number of which reside within Head Office, need to be identified and linked to a migration plan. However, while most officials in DWAF are aware of the CMA process, most are not prepared, in terms of what this means for their business area, and have not developed a strategy for the transfer of functions. This has impacts on business areas such as institutional oversight, water-use management, information management, resource-directed measures, finance, human resources, and the Regional Offices.

With such complex processes at work, it is easy for the process to be hijacked and stalled. While there is general Departmental support for the establishment of CMAs, it is also true that the establishment of CMAs has significant impacts on some people's roles and responsibilities. It is clear that some staff have felt threatened by this process, and in some instances, this has resulted in staff leaving the Department.

In retrospect, it is clear that there has been insufficient focus on an "enterprise" approach to the establishment and development of CMAs, and insufficient co-ordination of activities within the Department. The lack of a single, responsible figure in the Department to ensure that a comprehensive strategy was developed for the development of CMAs and the transfer of staff, funds and functions, allowed different units to pull in different directions and resulted in confusion, ineffective processes, and, in some cases, demoralisation of staff targeted for transfer.

7.5.6 Too Fast, Too Slow?

Linked to the "enterprise" approach that is required, there is a very real need to manage the periods under which the process takes place. In the processes so far, significant stakeholder fall-out has been experienced due to the drawn out periods. On the other hand, certain processes do need time. Clearly, these need to be understood and managed.

The institutional review process provides a very good opportunity to re-construct the establishment process and the required milestones and time-frames.

7.5.7 Transboundary Role?

To date there has not been significant effort placed on understanding the transboundary role of CMAs, despite the Inkomati CMA falling within a transboundary catchment. Since a large proportion of South Africa's water falls in transboundary basins (see Chapter 12), this issue is an important one that will need to be dealt with in the near future. Certain issues will continue to be dealt with at the national level, through inter-governmental agreements and relationships. However, implementation happens largely at the catchment level, and the CMA will have a critical role in this regard. Thus, it is important for the CMA to establish a relationship with similar structures in other riparian states in order to operate more effectively. The Inkomati CMA has already established a relationship with the Arasul, in Mozambique, and officials from each organisation have attended workshops and meetings offered by the other. This is an organic process, being driven by the CMA, but clear roles and responsibilities between the CMA and DWAF on international issues must still be determined, including the possible representation of the CMA on the transboundary structures.

7.6 Conclusion

While the policy on the establishment of CMAs remains appropriate for the management of water resources in South Africa, there are several areas in which implementation has been weak, and approaches need to be revised. The key areas to revisit have been highlighted in the section on lessons learned.

To date, much has been learned and will only assist the processes going forward. While these challenges are, in some ways unique to the South African context, many are also typical of any country going through a transformational process. What ultimately makes the difference is how one learns and responds to the challenges. It is, without doubt, that in any programme of institutional change, there will be discomfort.

Although the decentralisation-based reforms for water management defines a critical role for communities and users at large, the state must, and will continue to, play a fundamental role. This is because of its responsibility for managing water as a public good and for ensuring redress, equitable allocation of water and equitable representation in decision making. As a water-scarce country, the need to safeguard the sustainable provision of ecosystem goods and services and to protect the interests and welfare of all users, especially the poor, women and the disabled becomes especially critical (Molden, ed 2007). In working towards this objective, South Africa has recognised that catchment based management, together with the establishment of appropriate decentralised institutions, is an imperative. However, achieving this is not proving easy and is quite complex. Taking cognisance of lessons and reacting appropriately is important, as is the establishment and implementation of an "enterprise" approach to this process.

References

Asmal K (1998) Water as a metaphor for governance: issues in the management of water resources in Africa. Water Policy 1:95–101

De Coning C, Sherwill T (2004) An assessment of the water policy process in South Africa (1994 to 2003). Water research Commission Report TT232/04. Water Research Commission, Pretoria

DWAF (2005) The national water resource strategy for South Africa. Department of Water Affairs & Forestry, 2005. Government Printer, Pretoria

Molden D (2007) Water for food, water for life. Comprehensive assessment of water management in agriculture. IWMI, Sri Lanka

RSA (1997) White paper on a national water policy for South Africa. Government Printer, Pretoria

RSA (1998) National Water Act. Government Gazette Vol.398. 26 August 1998. Office of the President, Cape Town

Chapter 8
National Water Security: Planning and Implementation

Chris Moseki, Toriso Tlou, and Cornelius Ruiters

Abstract Since South Africa is a water scarce country, water resources planning is a critical part of ensuring national water security. Since 1994, there has been a paradigm shift to a more holistic approach to water resources planning, underpinned by South Africa's enabling Constitutional provisions, legislative framework and water policy, and implementation challenges. This new paradigm involves an integrated approach to water quality and quantity (for both surface water and groundwater) as well as consideration of water conservation and demand management, reuse/recycling, desalination and other options for water security at various levels of planning.

This chapter examines how planning for raw water supply contributes to ensuring equity in access for the livelihood of communities and to the economic growth and sustainable development of the country. As part of this, the alignment of water resource planning activities with provincial and local government needs, and some of the associated challenges, also forms part of the discussion. Infrastructure development, its role in water security and access, and related institutional arrangements are also addressed.

Keywords Water security • Water resources planning • Infrastructure development • Institutional arrangements • Integrated planning • Access to water • Equity • Sustainability • Water reallocation • Operation and maintenance • Climate change

C. Moseki (✉)
Water Research Commission, Pretoria, South Africa
e-mail: chrism@wrc.org.za

T. Tlou
Tlou Consulting, Pretoria, South Africa
e-mail: toriso@tlouconsult.co.za

C. Ruiters
Department of Water Affairs, Pretoria, South Africa
e-mail: ruitersc@dwa.gov.za

B. Schreiner and R. Hassan (eds.), *Transforming Water Management in South Africa: Designing and Implementing a New Policy Framework*, Global Issues in Water Policy 2,
DOI 10.1007/978-90-481-9367-7_8,

8.1 Introduction

South Africa is a semi-arid country in which the average rainfall is half the world average of about 860 mm per year (DWAF 2005). As a result, South Africa's water resources are, in global terms, scarce and limited in extent. Water is also unevenly distributed where certain areas are more stressed than others. Hence, making water available and accessible to competing users is a challenge to decision-makers and resource managers.

The situation was exacerbated, pre-1994, by the then legislative and policy position that favoured the minority, with emphasis on agriculture and commercial industry, while the majority of South Africans were marginalised and ignored. The unfortunate legacy of the country's notorious riparian access principle, discussed in Chapter 1, where the right of access to use of water was linked to land ownership, provides background to the present situation.

While South Africa is a water-scarce country, at the macro level there is sufficient water available for socio-economic development. This water is not, however, equitably distributed across the country and water surpluses do not match water demands from a spatial perspective. This has led to a complex, integrated system of cross-cutting infrastructure supporting economic and social development throughout the country. This infrastructure network, however, does not serve the rural poor to any great extent, and there is a challenge in ensuring equitable access, especially in cases of poor communities that need water for livelihoods (small scale agricultural and domestic use).

The challenges of water scarcity will be exacerbated by global warming and climate change which will increasingly affect water availability. High evaporation rates, resulting from raised temperatures will affect the availability of water in dams, rivers and perched aquifers, as well as on recharge and water quality. This often leads to incremental costs of building infrastructure to adapt to climate variability and change, and costs of creating climate information since historical data may be increasingly irrelevant in the calibration of models. To address some of these challenges, namely to achieve redress and equitable access to water, and to meet Reserve requirements, the Department initiated, developed and is presently implementing the Water Allocation Reform programme (see Chapter 5). The functions described above are supported by proactive water resource planning, project planning and design, as well as national infrastructure development, operation and management. However, a paradigm shift has been necessary in order to plan within the integrated water resources management (IWRM) approach adopted by South Africa.

The paradigm shift to a more holistic approach, underpinned by South Africa's enabling Constitutional provisions, legislative framework and water policy, and implementation challenges, is discussed in this chapter. This new paradigm involves an integrated approach to water quality and quantity (for both surface water and groundwater) as well as consideration of water conservation and demand management, reuse/recycling, desalination and other options for water security at various levels of planning.

This chapter examines how planning for raw water supply contributes to ensuring equity in access for the livelihoods of communities and to economic growth and sustainable development. As part of this, the alignment of water resource planning activities with provincial and local government needs, and some of the associated challenges, also forms part of the discussion. Infrastructure development, its role in water security and access, and related institutional arrangements are also addressed.

8.2 Access to Water

Under natural conditions, South Africa is characterised by low precipitation, high evaporation rates and a low ratio of rainfall to run-off as well as spatial and temporal variation in quantities of water available.

The lack of adequate availability of water resulting from natural climatic conditions is what Ohlsson and Turton (1999) refer to as first order water scarcity. Second order scarcity arises from the lack of infrastructure or management capacity to make available water accessible to people. South Africa faces a particular challenge related to historical allocation of water under the colonial and apartheid regimes. The resulting inequity in access to water has to be understood within the context of the socio-political history of this country (see Chapter 1).

As has been discussed in Chapter 1, the National Water Act (NWA) of 1998 is based on the principles of equity, efficiency and sustainability. It was intended to create an enabling environment for, among others, ensuring equitable access to water. Section 3 of the NWA entrusts the nation's water resources to the Minister of Water Affairs, acting on behalf of the national government to ensure management and control of water in a sustainable and equitable manner, for the benefit of all persons and in accordance with its constitutional mandate.

The National Water Act and the White Paper on a National Water Policy for South Africa changed the way water was to be developed, used, protected, conserved, controlled and managed. Based on the mandate of the NWA, the Department of Water Affairs and Forestry (DWAF) put in place a consultative process to develop the first National Water Resource Strategy (NWRS), published in January 2005. It is a long term national water resource master plan that broadly guides the Department, other organs of state, the private sector and civil society, on how water resources in South Africa will be developed, used, managed, conserved, controlled and protected in an equitable, efficient and sustainable manner for the benefit of all.

A key element of the NWRS is that it sets out the allocable water in each water management area. Some areas in the country are stressed in both quantity (over allocated) and quality (highly impacted) terms. To address this, both demand and supply side management are needed. As South Africa increases water storage for supply, water requirements and demands also have to be managed to ensure the sustainability of the resource. Lately, dams such as De Hoop in Limpopo and Berg River in the Western Cape have been built to increase capacity and make up for shortfall in those areas. However, prior to any consideration of the development of a water scheme, water conservation and demand management are to be taken into consideration. The Department wishes to be satisfied that water is used efficiently before venturing into dam building projects or initiatives that are aimed at increasing reservoir capacity or water supply. While this should be achieved by encouraging reuse of water, fixing bulk and reticulation networks, leakage control, using water saving devices, curbing irrigation water losses, proper irrigation scheduling, through awareness campaigns, supporting demand management projects and providing incentives to those who use water sparingly, more work needs to be done in this regard.

8.3 Integrated Water Resource Planning and Management

The NWA recognises that to achieve the objectives of sustainability, equity and efficiency, water resources need to be managed in an integrated manner (DWAF 2005). The White Paper on a National Water Policy for South Africa (1997) recognises the indivisibility of the hydrological cycle and stipulates that all water, wherever it occurs in the water cycle, should enjoy equal status. The IWRM approach, enshrined in the legislation, also requires the alignment of water, economic and social planning, an integrated approach to surface water and groundwater, and to quality and quantity aspects of water resources management and planning. Planning in an integrated manner marked a shift from previous practice where the focus, at the national level, was primarily on surface water and quantity elements.

The interaction between ground and surface water, in particular, has been a frequently discussed subject among researchers, managers and planners nationally and globally. Wherever interaction occurs, there is a likelihood of impact either in quantity terms where water that would, for instance, flow to the river is captured through pumping of boreholes, or in quality terms where contaminated river water impacts on groundwater resources (or vice versa). The interaction has significant implications for water resource planning and for water-use authorisation.

Under the Constitution of 1996, three spheres of government were established, each with its own set of functions assigned under the Constitution. In terms of managing the water resources of the country in a sustainable manner, the Department of Water Affairs and Forestry recognised the need to restructure itself to integrate surface water, groundwater and water quality under one directorate within the planning environment. Although organisationally there has been alignment, the key challenge facing DWAF is to mainstream integrated planning as part of its business processes.

The challenge to water resource planners is to ensure alignment with economic and development plans at all levels of government. Figure 8.1 shows the intended alignment between water and economic planning.

In order to support the alignment of economic and water planning, and in order to bring together, coherently, for the first time, information on each water management area (see Chapter 2), including surface and groundwater, quality and quantity issues, the Department developed a set of Internal Strategic Perspectives (ISPs). These were captured in the National Water Resource Strategy and provide key

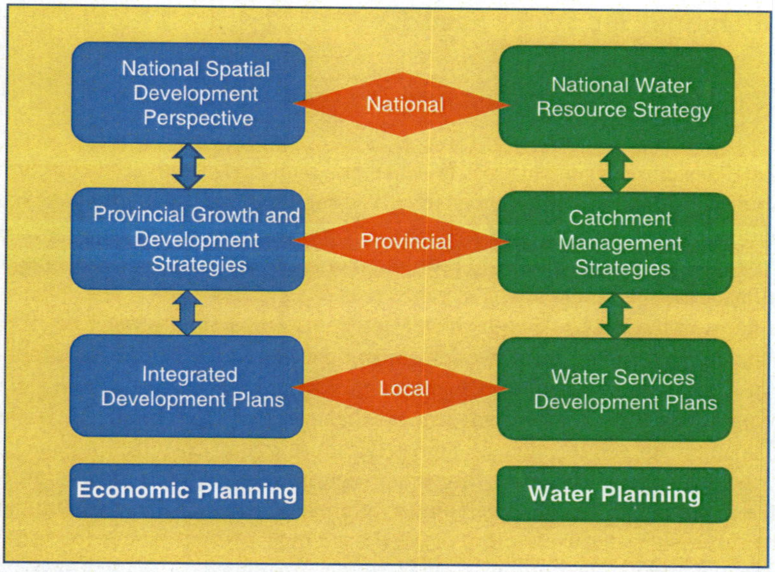

Fig. 8.1 Alignment of water and economic planning in three spheres of government

information on the water situation in each water management area. The ISPs provide first order strategies on water resource development and management within the water management areas and are the fore-runners for Catchment Management Strategies that, by law, should be developed by Catchment Management Agencies (CMAs) once they have been established. Unfortunately, however, the ISPs have not been used as widely as they should have been, including by DWAF water resource managers.

The ISPs should also inform the Provincial Growth and Development Strategies (PGDS) of provincial government, and the Integrated Development Plans and Water Services Development Plans (WSDPs) of the municipalities (see Chapter 2). These documents set out the economic and social development intentions of these two spheres of government. Unfortunately, it has taken a considerable amount of work to bring the alignment between water planning and economic planning. Alignment has been patchy, with some provinces and municipalities responding to the water challenges and constraints more effectively than others. For instance, the Mzimvubu Water Management Area (WMA) has been earmarked for development by the provincial government. However, where there is no sufficient water immediately available, DWAF supports the province in identifying alternative sources for poverty alleviation and growth initiatives, in the short term.

Overall, however, water resource management issues are not adequately addressed in WSDPs or PGDSs. DWAF has, as a result, recently embarked on a proactive process to plan, with municipalities, for their water needs. The process should take 3 years, beginning with the biggest municipalities and moving on to those facing the greatest water challenges (van Rooyen 2008).

8.4 Options for Water Security

8.4.1 Water Conservation and Demand Management

Of the 19 water management areas (WMAs) in South Africa, only 2 can be considered to have surplus water, while others are at varying levels of water stress. As has been mentioned, the Department highlighted water conservation and demand management as a national priority, inter alia, to prevent potential water shortages in the major metropolitan areas within the next 5–8 years. (van Rooyen 2008)

In most of the WMAs there is increasing competition among water users because current supplies cannot meet the demands at the required reliability of supply. This has, of necessity, forced a greater focus on improving water-use efficiency, and implementing water conservation and demand management. The approach has been to implement integrated resource planning (IRP)[1] using least cost planning methods.

One of the options to improving water security is the development of guidelines for the implementation of Water Conservation and Demand Management measures in the different water use sectors. DWAF has developed guidelines for WC/WDM for Water Services Institutions, (Water Services Authorities, Water Boards, Water Service Providers), the irrigation sector, and the Industry, Mining and Power Generation (IMP) sector.

The Water Research Commission (WRC) also developed various tools, models and techniques for developing and implementing WC/WDM measures.

8.4.2 Implementation Issues

Using these guidelines, WC/WDM Business Plans for various catchments were developed. These business plans are used in the development of reconciliation strategies to improve water security in the different sectors and catchments. Mainstreaming water conservation and demand management into planning and management has raised several institutional and regulatory issues that constrain effective implementation.

The organisational arrangements and existing capacity of Water Services Authorities, to ensure the implementation of water conservation and water demand management measures, is limited. Most metropolitan areas have some capacity to manage their demands but the poorer or smaller towns have challenges of dysfunctional and poor infrastructure, limited (if any) engineering capacity, and insufficient funds. The issue of budget is exacerbated by high levels of poverty in many of these areas (see Chapter 3).

[1] IRP is the process of meeting customers' needs for water use, considering a broad range of supply-side and WC/WDM options in a balanced manner. The objective is to secure the lowest long-term cost of water, consistent with the quantity and quality of water service desired by consumers (Swisher et al. 1997).

Recent proposals have been made within the Department for the establishment of a funding facility and implementation unit, at national level, that can assist in finding funding for water conservation and demand management initiatives, and provide technical support for the implementation of such projects.

8.4.3 Reuse of Effluent

Effluent reuse has not vigorously been pursued in South Africa, although some major industries, such as the pulp and paper industry and power generation, undertook significant effluent reuse programmes. This has, to some extent, been in response to the need to comply with international best practice and discharge standards, while in some water management areas it has been in response to water scarcity. At municipal level, the Durban metro is a particularly good example. Durban Water and Waste identified and undertook the following water demand management measures, including effluent reuse measures:

- Passive operational and maintenance measures on the distribution system, including a computerised control room to manage leaks, a centralised telemetry control of all reservoirs and pumps to minimise pressure fluctuations and spillages, and monitoring
- Proactive operational and maintenance measures on the distribution system, including pressure measurements, pipeline replacement, consumer metre management and leak detection
- Customer demand management measures such as block rate tariffs, repair of plumbing within households of poor communities and credit control measures
- Effluent return management and the development of a reclamation plant to treat 40 Ml/day and on-sell to industrial consumers

Another good example of efficient water reuse is the artificial recharge of groundwater by municipal return flows at Atlantis, near Cape Town. Municipal effluent is treated, injected underground for storage and, subsequently, used as required. This project has been in operation for over 20 years.

8.4.4 Groundwater Use as an Alternative Resource

Past neglect in the use and management of groundwater and groundwater's private status prior to the promulgation of the National Water Act (Act 36 of 1998), resulted in groundwater not being considered an integral part of the development and management of South Africa's water resources. Yet, groundwater is currently used to supply some 15 million South Africans with drinking water, and is widely used by the agricultural sector. This is clearly observable in the relatively dry areas of the Karoo where rivers are rare but windmill pumps are visible all over the place. The previously-ignored groundwater resource is recognised as a viable source of water, worthy of inclusion in integrated planning and management.

The WRC has undertaken several research projects on the Table Mountain Group and Karoo aquifers, including assessment of water availability and potential yield. However, many of the research outcomes still have to be implemented. DWAF has also done investigations regarding groundwater availability, with a Groundwater Information Programme (GRIP) conducted in the Limpopo Province, where a lot of data and information on groundwater was collected. The project is being replicated in the Eastern Cape and the Kwa-Zulu Natal regions. A national strategy for groundwater is also currently under development.

Groundwater is a useful form of storage, enhanced through artificial recharge of aquifers, since it is normally protected from evaporation and direct pollution. It is also a strategic resource that can be used to combat poverty and assist in meeting the Millennium Development Goals on access to water and food. For poor, rural communities, groundwater is often the only source of water that is readily available for livestock and to meet basic needs. The key challenge, however, is ensuring sustainable use and management, including preventative measures against pollution. Once contaminated, a groundwater resource is prohibitively expensive to rehabilitate and may never completely be rid of all pollutants.

8.4.5 Water Trading

The National Water Act provides for the trading of water either on a temporary or a permanent basis. Such trading is particularly active in stressed basins, between irrigation farmers, or between irrigation farmers and municipalities, mines or industry. Water trading serves to drive water use into the hands of more productive water users.

However, the South African context is more challenging than that of other water scarce countries because of the political and social imperative to introduce racial equity in access to water. Concerns have been raised in the Department that trading of water allocations is currently only between white users. The process of water trading results in the new user being issued with a water-use licence by the Department, while the previous user often only had authorisation under the "existing lawful use" clause of the National Water Act. Since a licence carries stronger legal authority than "existing lawful use" when reallocating water, the concern has been expressed that trading is entrenching water use in white hands.

The concern has also been expressed that trading has allowed water allocations to be moved out of areas where poor, black communities need water, into other provinces, once again, entrenching water use in white hands. The Department has, therefore, limited water trading opportunities until mechanisms can be found to resolve these issues and ensure that trading is used to promote transformation agenda and equity.

8.4.6 *Desalination*

Desalination of saline water is an option where no alternative water supply is available. Certain coastal towns in South Africa already have desalination plants. The process is quite expensive, but in view of water scarcity, it is also an option for water supply. High levels of salinity in the Vaal river system are largely the result of mining and industrial operations in the catchment. Deteriorating water quality is of serious concern, and the Department is considering desalination of mine water, in particular, to reduce salinity levels in this river system. This will, in turn, reduce the need for blending with good quality water and will, therefore, make more water available for use (van Rooyen 2008).

8.4.7 *Rainfall Enhancement*

Another option for water supply is weather modification. The first rainfall enhancement experiments in South Africa were undertaken in the 1970s, when a cloud-seeding experiment was launched in 1971, in the Bethlehem area. The aim was to increase rainfall in the Vaal Dam Catchment. Several research projects followed and continued until 1990 when weather modification research was amalgamated under the banner of the National Precipitation Research Programme that was first undertaken over a period of 5 years (from 1991 to 1995).

In 1995, the Limpopo Provincial Government approached the Water Research Commission, South African Weather Services and the Department of Water Affairs and Forestry, and requested the cloud seeding experiment to cover the Tzaneen area as an emergency response to drought. This culminated in the launching of the South African Rainfall Enhancement Programme (SAREP) in October 1997. SAREP was, initially, conceived as a 3-year project with joint-funding from DWAF, the Department of Environmental Affairs and Tourism and the National Department of Agriculture, and operated under a Policy Committee comprising Directors-General of these Departments and chaired by the WRC. During 1999, the Department of Water Affairs and Forestry initiated a number of studies in support of SAREP, to undertake environmental impact assessments and to determine and measure potential economic benefits.

However, the project has, over the years, lost momentum, partly because of a lack of sustainable funding from the parties. The South African Weather Services kept the technology alive and continued with the research, which culminated in South Africa winning a prize (equivalent to US $150,000) for this innovative technology in 2005, from the United Arab Emirates.

8.4.8 Removal of Alien Plants

South Africa suffers from high levels of alien vegetation invasion. In many cases, these alien plants use more water than the natural vegetation, particularly when growing in riparian zones. The removal of alien vegetation improves water availability and stream flow, as well as clearing arable land and restoring indigenous biodiversity. The world-renowned Working for Water programme, which combines job-creation in poor communities with the management and control of alien vegetation, has documented evidence of improved stream flow in areas where water-hungry alien vegetation has been removed.

In an innovative move, the Department issued the first licence to use water in a stressed catchment, based on the water user clearing and maintaining a specific area of invaded land. The water user is, in essence, removing the alien vegetation in order to free up water for use.

8.5 Infrastructure Development

South Africa has a proud record of accomplishment when it comes to the development of water resources infrastructure, and has the highest water storage per capita in Africa (Figure 8.2) – a fact that enables it to sustain lengthy droughts. The construction of this infrastructure facilitated economic growth, and still provides a reliable water supply for the country. However, the demands of further economic growth, population growth and improving standards of living, require additional water to be made available, including through the construction of more infrastructure.

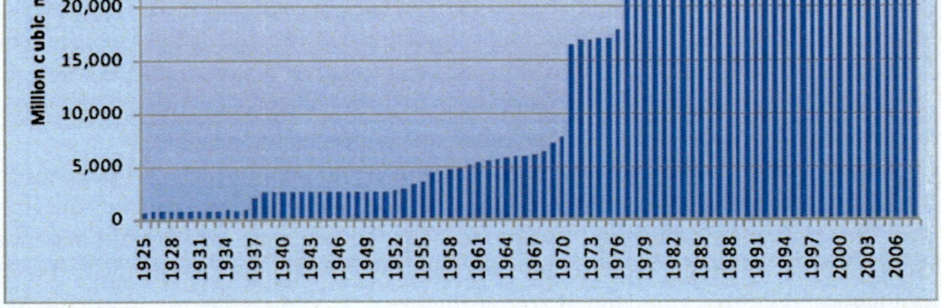

Fig. 8.2 Dam storage capacity 1925–2008 (Source Eales and Schreiner 2008)

The largest of the current developments is the De Hoop Dam in eastern Limpopo. When full, it will add an additional 347 million m^3 of storage capacity.

In terms of national infrastructure, DWAF has a comprehensive and advanced set of assets, including 218 dams listed in the ICOLD world register of large dams (dams with walls that exceed 15 m in height). If we look at the number of these dams built per decade in South Africa, it peaks at around 350 in the 1980s, and then drops sharply in the following decades. South Africa's water scarcity problem can no longer be solved by building more dams since the country does not have many appropriate dam sites where water is required. Alternative sources have to be pursued, such as groundwater development, rainwater harvesting and demand management.

In addition, many municipal systems, with associated housing and water services backlog programmes, as well as economic growth and development, depend on bulk infrastructure for water supply from major water resources infrastructure. There is a critical need to address the bulk water infrastructure needs since, for example, the infrastructure in more than 50% of the towns in some provinces, is limited and cannot support further growth. It requires urgent refurbishment and extensions with associated effective planning. In some areas, the introduction of water-borne sanitation, in place of the notorious bucket system, cannot be completed due to a lack of water availability.

Before 1994, the Department was largely responsible for the development of large water resources infrastructure, sometimes through the creation of special purpose vehicles. One such vehicle was the Trans Caledon Tunnel Authority (TCTA), established to provide funding for the Lesotho Highlands Water Project. This project was designed as an ambitious programme of dams and tunnels to provide water to South Africa from the Lesotho mountains. Since 1994, a policy of funding economic infrastructure off-budget saw increasing use of the TCTA, not only to fund, but also to implement, major infrastructure projects.

The importance of infrastructure development, operation and maintenance for the economy and social needs of South Africa is underscored by the increase in the national water resources infrastructure budget from around US $42 million in 2005/2006 to US $330 million in 2009/2010. The investment in infrastructure is highlighted by the fact that DWAF currently manages water resource infrastructure with an estimated depreciated replacement value of more than US $8.25 billion. In addition, South Africa's interest in the Lesotho Highlands Water Project (LHWP) is valued at US $2.1 billion. In co-operation with the TCTA implementation of further projects, with an estimated value of US $1 billion, has been initiated. The National Water Resource Strategy indicates that an estimated US $2.25 billion of water resources infrastructure will be required to meet the increasing water demand over the next 15 years.

8.5.1 Institutional Framework for Water Resources Infrastructure Development

With the promulgation of the NWA the arrangements for managing South Africa's water resources were set on a new course. Previously, three options existed for the development of large scale water resources infrastructure (mainly dams and large raw water conduits such as tunnels and canals):

- DWAF undertook this directly (by far the most common approach in the past).
- Water boards or large municipalities constructed their own facilities.
- Special purpose institutions were established (see above).

In order to provide for effective funding, development and management of water resources infrastructure in the future, the Department of Water Affairs and Forestry considered the establishment of a national water utility. The potential to create such an institution was included in the White Paper on a National Water Policy for South Africa. The option of creating more than one such utility, on a regional basis, was also investigated.

One of the key capabilities of such a utility would be that it should be able to raise the large amounts of capital required for the development of new, commercially viable infrastructure off the national government balance sheet. Current pricing policy requires commercially-viable infrastructure projects (or portions of projects) to be funded off-budget, while the social elements of projects are funded from the fiscus (see Chapter 9). The risk of financing such a body was assessed, primarily on the security of the income stream from sales of water from existing schemes, and those to be developed.

The development of a utility would also have allowed more market-related salaries to be offered to scarce occupational classes, enabling better attraction and retention of engineers, in particular. It would also have enabled the separation of regulator and regulated: currently, the Department operates and maintains dams, and is responsible for ensuring compliance with dam safety regulations; the Department also determines and regulates raw water tariffs.

The intention was that the DWAF infrastructure branch and the TCTA would be merged to form a national water resources infrastructure agency. Legislation supporting the establishment of the Agency was developed and, after several delays, presented to Parliament. Within Parliament, however, political changes within the ruling party and apparent concern regarding the impact of 'agentisation' on government workers led to the withdrawal of the Bill. For the near future, infrastructure will continue to be developed by the Department, and partly through the TCTA, while operation and maintenance and the setting of tariffs will remain the function of the Department.

8.5.2 National Water Resources Development Projects

The National Water Resources Strategy is the primary source of information on water availability. In compiling the NWRS, a major initiative was undertaken, specifically to assess where there are water resource deficits (both present and future), where there is surplus water and how these can be responded to. These figures have been updated for various water management areas since then, as necessary. The year 2007 picture is shown in Table 8.1.

Table 8.1 Reconciliation of water requirements and availability for year 2007 (million m³/a)

Water management area		Reliable[a] Local Yield	Transfers in	Local requirements	Transfers out	Balance
1	Limpopo	282	19	325	0	(24)
2	Luvuvhu/Letaba	310	0	334	13	(37)
3	Crocodile West and Marico	693	656	1,328	10	11
4	Olifants	611	172	971	8	(196)
5	Inkomati	943	0	1,048	148	(253)
6	Usutu to Mhlatuze	1,010	32	693	114	235
7	Thukela	738	0	338	497	(97)
8	Upper Vaal	1,723	1,443	1,204	1,481	481
9	Middle Vaal	201	791	389	605	(2)
10	Lower Vaal	50	651	653	0	48
11	Mvoti to Umzimkulu	527	34	828	0	(267)
12	Mzimvubu to Keiskamma	855	0	375	0	480
13	Upper Orange	4,557	2	968	3,105	486
14	Lower Orange	(1,007)	1,886	834	54	(9)
15	Fish to Tsitsikamma	437	571	902	0	106
16	Gouritz	277	0	342	1	(66)
17	Olifants/Doring	335	3	373	0	(35)
18	Breede	868	1	637	203	29
19	Berg	501	203	738	0	(34)
Total		13,911	0	13,280	124	504

[a] The amount that can be reliably provided for 98 years out of 100.

The National Water Resource Strategy acknowledges that more than half of the water management areas are in deficit (Table 8.1), although a surplus still exists for the country as a whole. The strategy suggests that the main interventions by which a balance between the availability of and requirements for water may be achieved are:

Demand management – The management of the demands (requirements) for water has been applied with great success by some users and could be used at water management area level to move from deficit to surplus. However, not enough has been done in this regard and the recent draft Strategy Framework for Water for Sustainable Growth and Development of the Department of Water Affairs identifies water conservation and demand management as a national priority.

Resource management – The regulation of streamflow through storage and the control of abstractions and releases, to provide appropriate quantities of water at specific times and locations (including inter-basin transfers), and of such quality and reliability to meet user requirements, is the traditional approach.

Development of surface water resources – There is.limited potential for further development of surface water resources in some parts of the country, through construction of storage and transfer infrastructure.

Inter-basin transfers – Inter-basin transfers are necessary to supply certain areas that experience water deficit, particularly Gauteng Province which is the industrial heartland of the country.

Developing and managing groundwater resources – Potential exists in dolomitic aquifers and the Table Mountain Group for larger-scale development of groundwater resources. There is, currently, considerable debate in the water community on just how extensive these resources are.

Reuse of water – While most water used in a non-consumptive manner is directly recycled for reuse or returned to the rivers for reuse elsewhere, there is further potential, particularly in coastal areas.

Control of invasive alien vegetation – Estimates suggest that close to 3% of the national mean annual runoff is intercepted by invasive alien vegetation. Removal and containment of such vegetation may contribute to enhanced water availability and improvement of biodiversity.

Reallocation of water – The reallocation of water between user sectors is an obvious and powerful option. The National Water Act provides for the reallocation of water to be introduced as the need arises in different parts of the country using pricing, compulsory licensing, water demand management, and trading of water-use authorisations.

Water quality considerations – Water quality is a fundamental concern in water resources management. In addition to making sufficient quantities of water available for use at specific locations and times, as required, reconciliation strategies must ensure that water is of appropriate quality for the intended uses. Currently, water quality problems in the Vaal river basin are partially managed by blending with good quality water. This reduces the volume of water available for other use.

8.5.3 Operations, Maintenance and Management

Based on its commitment to ensure effective management, DWAF initiated several studies, over the years, to investigate the condition of its infrastructure, including consideration of dam safety issues. Based on this study, the need to improve the condition of the water resources infrastructure was identified and resulted in special funding being allocated by the National Treasury. A backlog of maintenance and refurbishment had occurred, partly due to insufficient funds, partly due to the Department being the regulator and the implementer.

Infrastructure refurbishment is provided for on the Trading Account of DWAF. A Strategic Asset Management directorate focuses on the planning and implementation of major refurbishment and rehabilitation projects on existing infrastructure.

It was determined that the asset register of national water resources infrastructure was very weak, and an enterprise wide asset management project was launched to establish an effective asset register for water resources infrastructure. Detailed inspections are being conducted to establish the condition and remaining life of assets. This will further help the Department in prioritising refurbishment and rehabilitation plans, and in determining tariffs that are based on infrastructure value.

8.6 Lessons and Conclusions

Integrated water resources management requires planning that not only integrates surface and groundwater planning, and quantity and quality planning, but also integrates water resources planning into the economic and development planning of government. While some strides have been made in this regard in South Africa, there are still challenges in ensuring the alignment of planning between the various spheres of government and between the water and non-water sectors. Despite policy and legislative measures, such as the Intergovernmental Relations Framework Act, co-operative government has not been given proper effect to by the various spheres of government. Planning and developments have taken place in isolation, with minimal co-ordination and a lack of proper alignment between the various role players, despite the fact that water is an essential catalyst to all development. However, forums such as provincial Co-ordination Committees on Agricultural Water and the Provincial Liaison Committees have enabled DWAF to interact with other spheres of government on important water-related matters. Provincial and National Water Summits have raised the political profile of water resource issues among critical stakeholders.

The strong focus of government on addressing equity issues and the needs of the poor has also put pressure on DWAF to find ways to reallocate water and to address the needs of the poor. Pro-poor water resources planning is about prioritising access of water to the poor and considering poor communities' needs during planning.

A solid history of infrastructure development has placed South Africa on a sound footing for water security. There are, none-the-less, challenges to maintain this high level of security into the future. The challenges of looking, holistically, at the various options for reconciling supply and demand, including invasive alien plant control, water conservation and demand management and so on, have been driven as much by the policy and legislative framework as by increasing water scarcity. Limited surface water resources, the effects of climate change, economic growth and demographic pressures have encouraged DWAF to look to alternative resources and approaches. This has led to a paradigm shift regarding a more integrated approach in water management and planning. It has also led to more out-of-the-box thinking in terms of making water available as DWAF is now placing more focus on alternative and innovative approaches to dealing with impacts on water availability. The challenge, however, is successful implementation of the planning and policy shifts. Only with effective planning will the shifts result in increased water security in South Africa, for the poor, in particular.

References

DWAF (2005) The national water resource strategy for South Africa. Department of Water Affairs & Forestry, Government Printer, Pretoria

Ohlsson L, Turton AR (1999) The Turning of a Screw. Paper presented at the 9th Stockholm Water Symposium "Urban Stability through Integrated Water-Related Management", hosted on 9–12 August by the Stockholm International Water Institute (SIWI) in Sweden). Also available as MEWREW Occasional Paper No. 19. Retrieved 3 May 2007 from http://www.up.ac.za/academic/libarts/polsci/awiru

Swisher JN, Jannuzzi GM, Redlinger RY (2005) Tools and methods for integrated resource planning. Riso National Laboratory, UNEP, Denmark

Van Rooyen J (2008) Personal Communication. Director: National Water Resource Planning

Chapter 9
Pricing of Water for Cost Recovery, Economic Efficiency and Social Equity

Mahomed Vawda, Nicola King, and Mike Muller

Abstract The National Water Act (Act No.36 of 1998) recognises that water is a scarce and unevenly distributed national resource and that it is the responsibility of Government to ensure that the resource is managed in an equitable and sustainable manner. Due to the high costs and limited exploitable potential associated with supply-side water management solutions, demand-side management is becoming increasingly critical to ensure water security in South Africa. Fundamental to this approach is the appropriate pricing of water resources. The current water act aims to apply water pricing tiers based on the principles of economic efficiency, social equity, financial sustainability and ecological integrity. Failure to implement both supply-side and demand-side water management strategies effectively may result in the prospect of South Africa facing chronic water scarcity within 2–3 decades.

Keywords Supply-side management • Demand-side management • Water pricing • Economic efficiency • Social equity • Financial sustainability • Ecological integrity • Economic value • Opportunity cost • Stream-flow reduction

9.1 Introduction

The National Water Act (NWA) (RSA 1998) recognises that water is a scarce and unevenly distributed national resource and that it is the responsibility of Government

M. Vawda (✉)
Director: Water Resource–Finance and Pricing, Water and Environmental affairs, South Africa
e-mail: qac@dwaf.gov.za

N. King
Mintek, Private Bag X3015, Randburg, 2125, South Africa
e-mail: NicolaK@mintek.co.za

M. Muller
School of Public and Development, University of the Witwatersrand,
Private Bag 3 Wits 2050, Johannesburg, South Africa
e-mail: mikemuller1949@gmail.com; mike.muller@wits.ac.za

B. Schreiner and R. Hassan (eds.), *Transforming Water Management in South Africa: Designing and Implementing a New Policy Framework*, Global Issues in Water Policy 2,
DOI 10.1007/978-90-481-9367-7_9, © Springer Science+Business Media B.V. 2011

to ensure that the resource is managed in an equitable and sustainable manner (the policy for and regulation of water supply and sanitation services and their pricing is dealt with separately). Historically, water resource managers in South Africa met rising water demands through the establishment of a complex system of engineering supply-side solutions. However, due to their increasingly high costs and limited exploitable potential, these supply-side solutions are becoming less viable, and water managers are turning to the more attractive solutions offered by demand-side management (Priscoli 1998; Ashton and Seetal 2002; King 2002). Further pressures are now also being acknowledged by the establishment of ecological flow requirements for all types of water bodies, as set out by new NWA of South Africa, placing additional constraints on the availability of water for many off-channel users. South Africa may, possibly, face the harrowing prospect of chronic water scarcity within 2–3 decades (Smakhtin et al 2001).

This has obvious adverse implications, both for the national economy and the vulnerable, poor communities, which require various technical and food security interventions. It also needs interventions that offset market and institutional failures, while facilitating decision-making around water resource allocation and use for all South Africans. Intrinsic to this process, is the need to understand the true economic value of water resources beyond pricing measures that serve merely to recover costs in the absence of market prices and appropriate resource valuation systems. A poor understanding of the value of water resources will continue to encourage resource overuse and degradation, poor internalisation of the associated cost and benefit externalities of water use, and sub-optimal allocation among competing users, thus further compounding threats to development. The NWA makes provision for the use of economic instruments such as pricing, to manage water more efficiently and equitably. This chapter reviews the water pricing policies for South Africa. The following section provides an overview of the South African economy and water use. The third section discusses demand-side and supply-side options for water management. The evolution of water pricing policies and practices in South Africa prior to and after 1994 is reviewed in the following section. Practical examples of water pricing in different situations are presented, some funding and water pricing issues in South Africa are discussed and the chapter concludes with recommendations and considerations of future challenges.

9.2 The South African Economy and Water Use

South Africa is regarded as a water scarce country, with an annual per capita water availability of approximately 1,200 m^3 per capita per annum. Average annual rainfall is 497 mm compared to a world average of 860 mm, and this rainfall is unevenly distributed with 65% of the country receiving less than 500 mm annually and 21% receiving less than 200 mm. There are, also, frequent droughts that may last for several years. This picture is compounded by high annual potential evaporation of up to 3,000 mm in some areas. As a result of the country's topography and rainfall, 60% of river flow comes from 20% of the land area (DWAF 2004).

South Africa is a middle-income, developing country with an abundant supply of natural resources, and well-developed financial, legal, communications, energy and transport sectors. There is a modern infrastructure base supporting major urban centres and a rapidly growing economy – the stock exchange is tenth largest in the world – with extensive regional and global linkages. In terms of GDP, South Africa is in the top 30 countries of the world, with a GDP estimated at US $255 billion in 2006 and growing at between 4% and 5% (IMF 2007).

While a per capita GDP of $5,368 per annum in 2006 (IMF, 2007) places South Africa amongst middle income countries, it is a highly unequal society, with a Gini co-efficient (which measures the level of income inequality in a county) amongst the worst in the world. In 2006, it was estimated that over 40% of households were living in poverty (using US $2 as a measure), and 10% in extreme poverty (US $1 per day). Overall, official unemployment stood at 25.5% in 2006 (IMF 2007).

Although the agricultural sector contributes a small and declining share of total economic output, the sector is the most significant employer of labour in rural areas although there have been major job losses in agriculture over the last few years. While the primary agricultural sector is small, irrigated agriculture is the largest water user, accounting for 62% of total water use (Table 9.1).

Competition for water from other user sectors has resulted in declining availability of water for irrigation activities. However, any policy decision that affects the availability of water for irrigation has to consider the impact on labour as well as on the fact that, exports from secondary value adding activities of primary agriculture production, account for more than 15% of total exports.

Historic investment programmes have given South Africa a relatively well-developed water infrastructure, with inter-basin transfer schemes across the country and a network of dams giving a per capita storage considerably higher than for any other African country. This has allowed the country to maintain agricultural irrigation through dry periods. However, due to population growth and economic expansion, many catchments are facing water stress, with water demand exceeding supply (DWAF 2004).

The degree of potential water scarcity for South Africa is depicted in Fig. 9.1, which shows the total local water yield for 2000 and 2025, including baseline and high scenarios against the total local water requirements for the same period. Supply is estimated to remain relatively constant with minor adjustments for storage under construction in 2000, as reflected in the 2025 scenarios, and ranges from 13,911 million m^3 per annum (2000) to 14,681 million m^3 per annum (2025 baseline) and 15,460 million m^3 per annum (2025 high). Demand is expected to increase from 13,280 to 14,486 and 17,248 million m^3 per annum, under the baseline and high growth scenarios,

Table 9.1 South Africa's water use in 2000 (million m^3/a)

Irrigation	Urban	Rural	Mining/ industry	Power	Forest	Total
7,920	2,897	574	755	297	428	12,871
62%	23%	4%	6%	2%	3%	100%

Source: DWAF (2004).

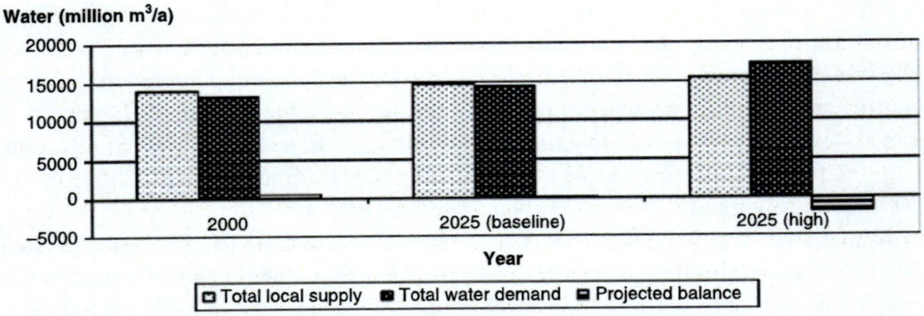

Fig. 9.1 Demand and Supply Balances for Water (over time) in South Africa, Excluding Transfers (Source: DWAF 2004)

respectively, potentially reducing the available surplus from 631 million m^3 per annum to a deficit of 1,788 million m^3 per annum under the high growth scenario.

At the household level, water supply and related sanitation services are also an important social issue, as reflected in the Millennium Declaration which sets a goal of halving the proportion of people without safe water supply by the year 2015. Sanitation was added by the World Summit on Sustainable Development held in Johannesburg in September 2002. In 1994, one-third of the people in South Africa did not have access to safe water and half did not have hygienic sanitation facilities.

This heightened awareness of the scarcity value of water resources and its impending consequences has encouraged water managers and ultimately users to recognise that the value generated by, or dependent on, water for the provision of goods and services, recreation, life support and ecosystem functioning must be determined and well understood (Ashton 2002). As a result, government interventions through demand or supply side management are necessary to address the gaps in delivery and access to water in South Africa, while at the same time being mindful of the increasing poverty gap.

One intervention is that of water pricing and tariff setting. This is a complex and controversial tool for water management, but the new NWA introduces some innovative ways of meeting water demand through pricing, which are outlined later in this chapter.

9.3 Funding and Pricing of Water and Its Services in South Africa

Appropriate pricing of water and water services has become an important issue for South Africa and will have substantial impacts on demand. For instance, if the price of water is set high, farmers will be induced to move to more efficient forms of production (if they can afford the necessary investments) but if set too high, they might abandon production altogether. On the other hand, if the price is too low, that will provide an incentive for overuse and wastage of water, and not enough funds will be collected to maintain or expand the supply infrastructure. Other economic users of water may behave in the same way although water is rarely as important a factor of production for other sectors. Similarly, if the price for safe water supply and sanitation services in poor communities is too high, many households will ignore the improved infrastructure and resort to unsafe options. If the price is too low, maintenance and expansion of water infrastructure will not be undertaken, leaving other households inadequately served while the better served (who are usually the better-off) benefit from lower prices.

In more "normal" goods and services markets, prices would typically be determined by the private market to recover capital, operation and maintenance costs of delivering the service, plus a margin of normal profit returns, on invested capital[1]. However, the socio-economic importance of water-dependent sectors such as agriculture and the "human right" or "public good" nature of basic water services makes water an atypical private good as there are clear reasons for intervening in the market setting of prices for water in at least some of its uses.

Where the benefit from water-use contributes to a public good (such as better health for the entire community) or general social welfare goals (such as increased employment in rural areas), there may be sound reasons for, at least, part of the costs of the water supply or service to be funded from other sources, usually general taxation (or, in the case of large systems covering a range of communities and activities, by cross-subsidisation from one group of users to another). It is useful, in this context, to consider the funding of water-related infrastructure and services separate from the pricing of those services. These considerations have, to a greater or lesser extent, guided the approach to water pricing in South Africa.

9.3.1 Water Resource Pricing Before 1994

As has been shown elsewhere in this volume, institutional responsibility for the management of water initially focused on the needs of agriculture and the present DWAF had its origins in the Department of Irrigation of the new Union Government which came to power in 1910. Before 1994, water management served the political and economic interest of the minority ruling class, the distribution of water-use was racially biased, and the majority of South Africans were denied direct access to water and the benefits from its productive use. Meanwhile, heavily subsidised water schemes were used to address the political challenges of the white community.

However, pressures of water scarcity were growing and the 1970 Commission of Enquiry into Water Matters (DWA 1970) reported "exceptionally high subsidisation of irrigation schemes, and water rates that are amongst the lowest in the world" and recommended the adoption of a realistic price policy to reflect the scarcity value of

[1]Note that the resource rent which reflects the scarcity value of the resource (such as water) will accrue in full or in part to either the provider (the water utility), government or users depending on property rights systems and production relations in place.

relevant factors. The Commission accordingly made the radical (for the time) proposal that; "… water from Government water schemes should be supplied to all consumers, whether for consumption or processing, at a tariff such that after taking account of all direct, indirect and intangible benefits, as well as assurance of supply, is sufficient to cover capital and operating costs" (DWA 1970.)

This implied that water rates on existing Government irrigation schemes be gradually raised to cover at least the operating costs. A Committee of Enquiry into the Price Policy with Regard to the Determination of Water Tariffs, appointed in December 1980 (DWA 1984), found that:

- Over a 6 year period, tariffs for irrigation rose by 120% but remained below the full operating cost of Government water schemes.
- Small farms could not absorb further water price increases without suffering a decline in their standard of living.
- In view of the vast difference between the industrial and agricultural sectors, a differential policy of tariff determination for these two sectors was justifiable.

As a result, a revised raw water price policy was established which diluted many of the original recommendations insofar as the farmers were concerned but retained some of the more far-reaching basic principles. For example, Government decided that full costs recovery on water supplied for irrigation and stock watering should not form the basis for tariff determination. Rather, the determination of tariffs should be based on how much the average farmer on each scheme could afford to pay for water. However, it was decided to retain the existing policy for domestic and industrial consumption of water, implying that the tariff would be levied to cover the cost of the supply of water including capital cost. In developing new domestic and industrial water supply schemes, it was decided to apply a marginal-cost approach in tariff determination (i.e. the cost of providing an additional cubic metre of water) to reflect the cost of augmenting existing supplies.

During this period, the requirements of industrial and urban users did not receive much attention since conurbations, such as Johannesburg and the Witwatersrand area, Durban and Cape Town, were able to develop their own water supplies and local water pricing was considered to be a municipal affair or, where there was a need for a regional approach to water supply, undertaken by multi-jurisdictional water boards. The Commission (DWA 1970) found that water was rarely a significant cost to industry, representing on average just 0.17% of the cost of all materials in the manufacturing industry. Some subsidies were made available for municipalities facing particularly expensive water supply costs and the Commission felt that, in the national interest, sewerage purification should be subsidised, requiring local authorities to make the water available for re-use.

9.3.2 *Water Pricing After 1994*

However dramatic the political transformation that occurred in 1994, the underlying physical realities – and the policy conclusions that resulted from them – did not change substantially. What did change was the relative weight of different interest groups. While

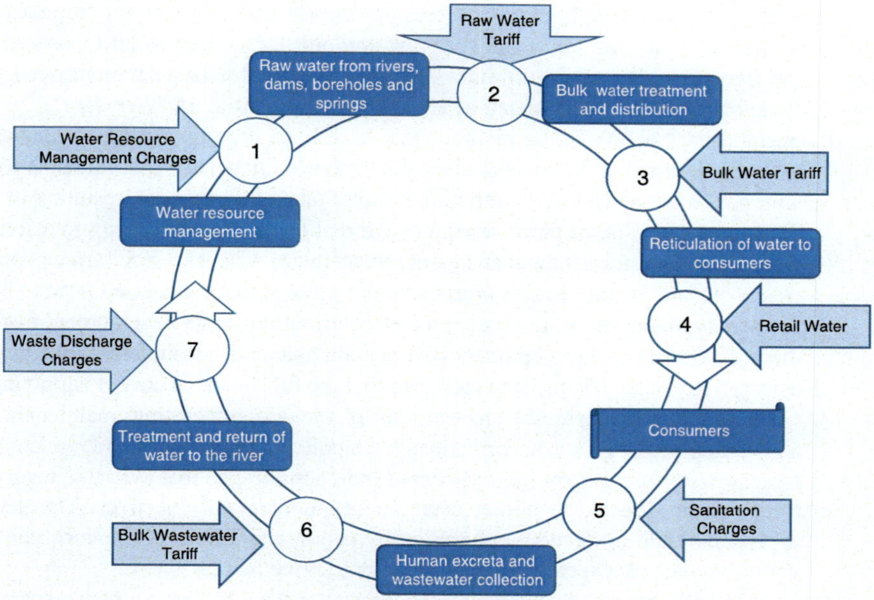

Fig. 9.2 Water cost and pricing chain (Source: DWAF 2003)

before 1994, the interests of a small number of white farmers were dominant, after 1994, attention shifted to the needs of the poor, black majority (Muller 2007). To fulfil the Constitutional mandate, a review of the national water-use policy and water law was required by the post-1994 Government. This Constitutional right requires equitable access by all South Africans to the benefits of the national water resources and an end to discrimination on the basis of race, class and gender. The National Water Policy (RSA 1997) and National Water Act (NWA 1998) provided the basis for a new pricing strategy for water resources, amended in 1999 (DWAF 1999) and recently in 2007 (DWAF Deparment of Water Affairs and Forestry) (2007) Establishment. of a princing strategy for water use charges, Goverment Gazette No. 25697 16 March 2007 The new water pricing strategy formalised the dis-aggregation of pricing regimes into pricing tiers, namely raw water, bulk water and retail water pricing components (Fig. 9.2).

Since water supply and sanitation services are, constitutionally, a local government function, a separate approach to their pricing was developed under the Water Services Act (DWAF 1997) within the broader framework of municipal financing policy.

9.3.3 Objectives of the Post-1994 Water Pricing Strategies

The pricing of water was explicitly recognised as a key element of the water demand management strategy, and the overall strategy was guided by four key objectives:

Social Equity – This is to be achieved through the provision of financial assistance to redress the imbalances of the past, both with respect to equitable access to water supply services (retail/domestic water) and with respect to direct access to first tier water (i.e. raw and bulk). One important measure is the free provision of a minimum amount of water for basic human need – "free basic water" – estimated at 25 l per person per day (DWAF 2002).

Financial Sustainability – The methods that have been used by DWAF to finance major bulk raw (i.e. first tier) made water schemes financially unsustainable for a number of reasons. Firstly, inflation was not taken into account, resulting in a decline in the value of tariffs over time, in real terms. Secondly, no provisions were made for refurbishment and asset replacement. A new financial framework was, therefore, required to accommodate the water sector's increased need to be financially autonomous, to attract greater contributions to its development from the private sector, and to be financially accountable and sustainable. In the new approach to water pricing, it is proposed that the full financial cost of supplying water should be recovered from water users, including operating, maintenance and capital costs. This is accommodates the policy requirement that large commercial water users carry the full cost of their supplies, so that available investment can be directed to poorer users and explicit social objectives. The new approach would however, be phased in by taking cognisance of the constraints within various user sectors to adapt quickly to price increases.

Economic Efficiency – Ensuring an efficient allocation of the country's scarce water resources requires that the price of the resource be set to reflect its scarcity in those areas where water is already fully utilised. In such cases, failing to price water at its scarcity value can result in over-use, leading to premature expansion of infrastructure. Moreover, not charging the scarcity value encourages use of some water in low-value activities. This imposes an opportunity cost as this same water cannot be used for alternative, high-value purposes. In the context of water scarcity, an argument can be made for the introduction of economic incentives in water-stressed catchments to encourage the conservation of water and its shift from low to higher value use.

Ecological Sustainability – This is based on the principle that there is a cost associated with the ecological management of the catchment, which should be paid for by all the users of the resource. To preserve water quality, point and diffuse sources of pollution should be discouraged through the identification of more effective control methods and instruments such as the "polluter pays" approach, such as through a waste discharge charge.

9.3.4 Types of Charges Under the New Water Resource Pricing Strategy

Under the 1999 strategy, all significant water resource users, including forest plantations (defined as a stream flow reduction activity) are charged. The only exceptions as free basic rights are the reserve for basic human needs, the ecological reserve and small abstractions undertaken within the framework of 'general authorisations'.

Further, the price of water varies according to location and is calculated on a system, catchment or sub-catchment basis, including operating, maintenance and capital costs, as well as a water resource management charge (WRMC). The WRMC is determined to recover the cost of water resource management incurred by the relevant administration in each water management area (WMA) in the country, and is payable by all water users (excluding those defined in the National Water Act as Schedule 1 users).

The WRMC is based on the cost of providing a water management service and relates to those activities that are required to regulate, manage and maintain the water resource or catchment. These costs are not related to water sales but represent all costs of management within a WMA including: planning and implementation of catchment management strategies; monitoring and assessment of water resource availability and quality; flood and drought management; management of raw water distribution, control, storage and stream flow reduction; evaluation and processing of water-use licensing and registration applications; water resource protection, conservation and pollution control. While water resource management is currently the task of DWAF, Catchment Management Agencies (CMAs) will eventually take over these functions and activities, which will be funded from the WRMC. The WRMC is levied on registered use and was capped in 1999 at a maximum of 2 cents per cubic metre, increasing with inflation, to avoid the creation of bloated CMA bureaucracies.

The new water pricing system includes another charge known as the water resource development costs, which can include the related costs of investigation, planning, design and construction of water schemes, constituting the capital cost of projects. These are in addition to operations and maintenance charges on government water schemes and they include: direct costs representing the fixed and variable costs of administrating, operating and maintaining schemes, pumping and distribution of water; and indirect costs that cannot be directly attributed to a specific scheme, but which contribute towards the management of the water schemes of the entire area and comprise of allocated regional and area office costs.

In addition, the new pricing system collects a return on assets charge based on a fair rate of return to capital employed by Government to finance the development of water infrastructure. This rate is set at 4% of the depreciated value of the asset. This component of the charge is set on a scheme-related basis and is only applied to those sectors with increasing demands. These include local government, industry, mining and energy but exclude agriculture.

The 2007 pricing strategy extended the system by proposing a waste discharge charge system and the financing of infrastructure with off-budget (non state) funding.

Another charge is provided for in the strategy to be applied in areas where there is competition for water. This so-called economic charge, is intended to support the objective of economic efficiency by providing an incentive to shift water use from low to high value use. The economic charge can be set only by DWAF, on a scheme or system basis, either administratively, by determining the economic value of water, or via market-based mechanisms. Administrative mechanisms may be used in water stressed catchments based on the opportunity cost of water as determined by prevailing trading transactions. The administratively set economic charge will not be introduced before compulsory licensing is implemented, and then only after consulting the relevant stakeholders and water management institutions. This annual charge will be an add-on to any water charges. Market-based mechanisms will be applied where water is still available for allocation after compulsory licenses have been issued and there is competition for using this water. A public auction procedure may be followed for this purpose where applicants are allowed to take up the available supply through bidding or tendering process. A market-based mechanism which leads to transfer of water use entitlements via trading transactions is already in place and procedural guideline have been developed to facilitate trading within and between water use sectors with limited controls on transactions within a sector in the same area but more onerous controls over transfers between different sectors and regions.

9.4 Examples of Water Pricing in South Africa Post-1994

This section reviews the actual practice of water pricing in South Africa under the NWA.

9.4.1 Raw Water Pricing

The NWA requires that water pricing be established in terms of a formal Pricing Strategy which "refers to pricing the use of water from South Africa's water resources, and not to the pricing of water services". Water services and their pricing are regulated separately by the Water Services Act (DWAF 1997). Raw water pricing focuses on the use of raw water from the water resource (Fig. 9.2). It only affects second and third tier water (i.e. water supplied in bulk often by water boards) and distributed to households (usually via a water services authority) as an input charge for those services.

In its introduction, the 1999 Strategy reflected the theme already raised by the Commission in 1970: "…it is important to make sure that the water that is already available is used efficiently and not wasted. This is best achieved by introducing demand-side measures to manage our water resources. By encouraging all water sectors to use water more efficiently, demand management provides a more sustainable long-term solution to the problem of water scarcity than do supply-side measures, because it takes into account the value of water in relation to its cost of provision, thereby treating it more like a commodity."

9.4.2 Bulk Water Pricing

Bulk water is provided by Water Boards which abstract, treat and transport bulk water to service providers like municipalities, or by municipalities themselves. As a secondary activity, some Water Boards also undertake retail water supply on

Table 9.2 Summary of raw water charges in the 2005 pricing strategy

Sector	Resource management charges	Resource development charges	Phasing in of charges
Domestic/industrial	Full cost recovery on abstraction and waste discharge related use	GWS: Depreciation; ROA: O&M WMI's: Full cost recovery	WRM charges introduced fully after registration of water use in WMA Waste discharge related WRM charges to be implemented in 2007/2008 Annual increase on development charge will be limited to PPI + 10% until target development charge is achieved on GWS
Stream flow reduction activities: Commercial growers	Full recovery of allocated costs Note: Cost of dam safety control and waste discharge related costs not allocated to the forestry sector.	Not applicable, except where negotiated for new development	WRM charges introduced fully after registration but capped to R10 per ha plus PPI with 2002/2003 as base year
Stream flow reduction activities: Resource poor growers	Full recovery of allocated costs to be achieved in 5 years Note: Cost of dam safety control and waste dischargerelated costs not allocated to the forestry sector.	Not applicable, except where negotiated for new development	As above, but subsidised for 5 years from date of registration Subsidy starts at 100% and reduces by 20% annually No charge for forest plantation that is ≤ 10 ha
Irrigation: Commercial farmers	Full recovery of allocated costs Waste discharge related costs not applicable	GWS: Full recovery of Depreciation plus O&M on existing schemes Full financial cost recovery for new schemes WMI: Full financial cost recovery	Depreciation charge capped to 1.5 c per m^3 plus PPI WRM charge introduced fully after registration of water use in WMA, but capped to 1.5 c per m^3 plus PPI from 2006/2007
Irrigation: Resource poor farmers	As above, but subsidised for a 5-year period Waste discharge related costs not applicable	GWS: O&M subsidised for a 5-year period on existing and new schemes Depreciation charges waived for a 5-year period WMIs: Subsidies available under certain conditions	GWS: O&M charges phased in over 5 Years after registration at 20% per annum Depreciation charge applied from year 6 onwards and capped to 1.5 c per m^3 plus PPI from 2006/2007 WRM charge phased in over 5 years at 20% per annum

behalf of poorly capacitated municipalities and operate bulk wastewater works. Raw water drawn from dams and other schemes has to be purified and distributed to service providers. Where water boards perform this role the cost incurred is covered through "bulk water tariffs", which are subject to consultation with customers and stakeholders in terms of Section 42 of the Municipal Finance Management Act.

Given the very different circumstances of the different Boards, there is no uniform method in setting bulk water tariffs, although the Water Services Act (DWAF 1997) sets the criteria for Water Boards as follows:

34. (1) In performing its activities, exercising its powers and carrying out its duties a Water Board must achieve a balance between:

(a) striving to provide efficient, reliable and sustainable water services;
(b) optimally using available resources;
(c) striving to be financially viable;...

For the purpose of subsection (1)(c) a Water Board is financially viable if it is able to:

(a) repay and service its debts;
(b) recover its capital, operational and maintenance costs;
(c) make reasonable provision for depreciation of assets;
(d) recover the costs associated with the repayment of capital from revenues (including subsidies) over time; and
(e) make reasonable provision for future capital requirements and expansion."

This is more specific than the water services tariff regulations (DWAF 2002) which also govern municipalities as water service providers, which simply provide that:

A water services institution must, when determining its revenue requirements on which tariffs for water services are based, take into account at least the need to:

(a) recover the cost of water purchases;
(b) recover overhead, operational and maintenance costs;
(c) recover the cost of capital not financed through any grant, subsidy or donation;
(d) provide for the replacement, refurbishment and extension of water services works; and
(e) ensure that all households have access to basic water supply and basic sanitation.

In terms of municipal financial regulations, the Boards are also required to discuss and agree with municipalities on their prices. This has gone some way to promote transparency. While no explicit policies exist a draft guideline has been developed by DWAF. Charges are generally set at a cost-plus price but there is no formal economic regulation of these prices, and no clear guidelines for the allowed costs or rate of return have been established. As a result there are limited incentives to reduce costs or improve efficiency although DWAF maintains a policy that increases should, where possible, be consistent with government inflation targets.

Box 9.1 The Umgeni water example of bulk water tariffs

Umgeni Water's tariff model uses various base case assumptions for future water demand, inflation and interest rates and debtors payment periods. It also takes into account the expected lifespan of the utility's assets by providing for capital expenditure projections, as well as all loan repayments and financing of debt. The tariff for water charged by Umgeni Water to its consumers is made up of the following:

- The raw water charges as charged by DWAF.
- The direct operating costs as incurred by Umgeni Water in delivering bulk water.
- The administration costs as incurred by Umgeni Water in delivering bulk water.
- The net finance costs related to the cost of borrowings to finance the infrastructure of waterworks, reticulation schemes and working capital requirements.

Source: Umgeni Water (2005)

The five largest Water Boards generate some 91% of all Water Boards revenues, with Rand Water (US $600 million) and Umgeni (US $175 million) by far the largest. The value of their combined assets amounts to about US $1.3 billion, with an external debt of just under US $ 800 million. Capital spending among the 15 Water Boards varies, with Rand Water and Umgeni making capital investments during 2007/2008 of around US $60 million and US $26 million, respectively (Box 9.1).

9.4.3 Retail and Domestic Water Pricing

To help redress past discrimination, the 1996 Constitution placed a duty on all three spheres of government (national, provincial and local) to realise the right of access to water by acting in partnership with one another. The Constitution itself required that: "A local government shall … make provision for access by all persons residing within its area of jurisdiction to water, sanitation … within a safe and healthy environment, provided that such services and amenities can be rendered in a sustainable manner and are financially and physically practicable" (RSA 1996).

The responsibility for water services rests primarily with municipalities as established under the Water Services Act (DWAF 1997). The Act assigns municipalities with the water services authority function. The actual delivery of services is undertaken by water services providers appointed by the water services authority. Whilst most municipalities are water service providers, some municipalities utilise municipal entities (for example, Johannesburg Water), water boards or contracted service providers as their water service providers (see Chapter 3).

There is no country wide standard format applied by municipalities in pricing potable/retail water. Water service providers are constrained by inflation targets, and social, political, economic and consumer interests in pricing retail water. These factors make it difficult for municipalities to pass all of their increases in input cost on to consumers. While prices may be set administratively, the discretion of politicians in setting retail water prices severely constrains full cost recovery. This often results in inadequate investment in and maintenance of local water infrastructure.

The Municipal Systems Act of 2000 (RSA 2000) contained two key principles. Firstly, local government must aim to provide broad access to basic services and should make use of cross-subsidisation to achieve this goal. Secondly, local government must recover costs when delivering services to ensure that it remains financially sustainable.

To provide a practical guide to address this potentially contradictory position, regulations were promulgated in terms of the Water Services Act (DWAF 2002) requiring municipalities to adopt a "rising block tariff" system. In terms of this, the cost of water increases with usage, subject to the requirement that the first block, up to 6 kl per household per month should be provided free wherever feasible. The price then increases with the volume used so that those who use large amounts of water pay more than the unit cost and contribute to the cost of the "free basic water". The higher tariff blocks also serve to incentivise conservation. In addition to user charges, municipalities receive an "equitable share of national revenue" calculated to cover the operational costs of water supply to poor households, enabling the system to function in municipalities without sufficient high income large users to support a cross-subsidy.

This approach, adopted in 2001 when the new local government system was established, went well beyond the initial policy outlined in the ANC's Reconstruction and Development Programme (RDP) (African National Congress 1994) which required a tariff structure with a lifeline tariff to ensure that all South Africans are able to afford water services sufficient for health and hygiene requirements; and a progressive block tariff in urban areas to ensure that the long-term costs of supplying large-volume users are met and that there is a cross-subsidy to promote affordability for the poor.

Free basic water is now implemented by most South African municipalities, to differing degrees (see Chapter 3). The policy has been controversial and there have been complaints that it is not sustainable and is poorly targeted. However, analysis suggests that this is only the case where other elements of the tariff have not been appropriately set or where budget transfers (the equitable share of revenue) intended to enable basic services to be provided to the poor, are used for other purposes. (Muller 2008).

One problem with the implementation of stepped tariffs is that if water consumption were to go down, then all prices would have to go up, because the same income is required from water consumers to service loans for existing infrastructure. This therefore becomes a disincentive to implement demand side measures to conserve water. In order to mitigate such a situation, municipalities will have to include some 'fat' in revenue collection to anticipate an overall reduction in consumption, which could come about by measures to reduce household water losses.

9.5 Other Contemporary Funding and Water Pricing Issues

9.5.1 New Infrastructure Development and Pricing

DWAF is the custodian of some 300 large dams throughout South Africa. The National Water Resources Strategy (DWAF 2004) sets out potential future national infrastructure for South Africa's future water needs with an estimated investment in excess of R20 billion over a period of 20 years for dams and related projects. While some schemes are financed with state funding the NWA allows the Minster to implement and fund commercially viable schemes with non-Governmental funding. The Trans Caledon Tunnel Authority (TCTA) is a State owned specialised liability management entity responsible for financing raw water infrastructure development in the most cost-effective manner for the benefit of water consumers. This allows TCTA to raise loans to finance the development of new infrastructure on the understanding that the loans will be serviced through cost recovery from economic water users. The TCTA in consultation with stakeholders can determine, on a project-by-project basis, the extent of charges as determined by the proposed financial modelling. The primary charge is usually the capital unit charge (CUC) which should be sufficient to repay the loan over an agreed period. Off-budget funding of schemes will be applied in instances where schemes are developed for commercial use or for domestic use by communities that can afford the resulting tariffs.

The Berg River project, commissioned in 2008 to augment Cape Town's water supply was financed on this basis as was the US $330 million Vaal River Eastern sub-system augmentation project, which involves the construction of a 116 km pipeline to channel 160 million m^3 of water a year from the Vaal River to Secunda. The TCTA financed the project with money raised from capital markets which will be recovered in tariff charges from the beneficiaries, Eskom and Sasol.

The CUC tariff is determined such that the debt the debt repayment period is not longer than the economic life of the asset; on the basis of the debt profile and acceptable growth and level of debt of the project; not allowing the debt of a project to overlap to an unreasonable extent with another project causing financial strain to end users or unhealthy financial balance in the water sector; and taking into account the anticipated future funding requirements of the augmentation of the project.

The CUC tariff is calculated such that it is constant in real terms unless otherwise agreed to between the parties and it may be phased-in during the construction period. Parties should endeavour not to capitalise interest after completion of construction. Demand projections are used to determine a CUC tariff which is reviewed annually taking account of changes in: water demand projections; real interest rate projections; inflation projections and/or cost of the scheme as well as cost and timing of future augmentation.

9.5.2 The National Water Resource Infrastructure Agency (NWRIA)

In late 2005, Cabinet approved the establishment of a NWRIA, aimed at ensuring long-term water security for South Africa. The agency was intended to develop and operate South Africa's major national dams and water-transfer schemes, which are currently managed directly by DWAF. These assets are estimated at approximately US $8 billion at depreciated replacement cost, and bulk water sales bring in more than US $270 million annually. The NWRIA was to incorporate the Trans-Caledon Tunnel Authority (TCTA) and the infrastructure branch of DWAF. DWAF would retain the policy, oversight and regulatory work of water resources management.

The proposed agency would, based on its large asset base, be able to leverage finance from capital markets for water resources infrastructure. It would also be able to make a "social dividend" available to finance social infrastructure to meet basic human needs (domestic) and redress past wrongs. It would also reduce the dependency on fiscal appropriations. However, changes in national institutional policy and complaints from organised labour that they had not been adequately consulted on the proposals have resulted in the stalling of the establishment of the agency.

9.5.3 Water for Irrigation and Subsidies for Poor Farmers

Agriculture forms a key, if small, part of the South African economy, providing a crucial source of employment in rural areas. It is also the largest water use sector in the country. Agricultural development under colonialism and apartheid saw the creation of a relatively strong white commercial agricultural sector and the destruction of the once vibrant black farming community. In South Africa, as in many parts of the world, agriculture is under considerable pressure from increasing input costs and decreasing profits. It is also vulnerable to long droughts, which may well be exacerbated by global climate change.

To support new black farmers with limited resources, DWAF developed a policy to provide financial assistance in terms of Section 61 of the NWA with funds appropriated by parliament. The assistance covers:

- Capital cost for the construction and/or upgrading of off-farm infrastructure for irrigation schemes.
- Subsidies for Operation and Maintenance (O&M), Water Resource Management (WRM) and Depreciation Charges.

- Grant support for buying water allocations (water trading). In many areas of the country all available water has already been allocated and the only means of access to water is through trading of allocations.
- Grant support for socio-economic viability studies and investigations and for training of Water User Association (WUA) members.
- Financial support for rainwater harvesting.

This support aims to promote economic development in rural areas where there are few development opportunities. In all of the above-mentioned cases except for subsidies on water charges and rainwater harvesting tanks, grants/subsidies are made available to the relevant WUA or approved legal entity. A higher level of financial support is provided to WUA management committees where the proportion of women in the management committee is equal to or higher than the proportion of local women farmers. Funds are provided from the annual budget of DWAF. Additional support is also made available (US $33 million in 2005/2006) by the Department of Agriculture for small-scale infield agricultural development such as acquiring farm implements, including tractors, infield irrigation equipment and for working capital like acquisition of seeds.

9.5.4 Waste Water Management and the Role of the Water Discharge Charge System

Section 56 (5) of the NWA enables the Minister to establish a system for charging waste discharges in terms of the pricing strategy. The waste discharge charge system (WDCS) is based on the polluter pays principle and aims to:

- Promote the sustainable development and efficient use of water resources
- Promote the internalisation of environmental costs by waste dischargers
- Recover costs associated with mitigating resource quality impacts of waste discharge
- Create financial incentives for waste dischargers to reduce waste and use water resources in a more optimal manner

The WDCS may be implemented in catchments for which resource quality objectives (RQO) are either exceeded or are threatened (see Chapter 6). The WDCS will be applied at a catchment scale, not on a WMA scale. This may therefore be an entire catchment in which a widespread water quality problem exists or a sub-catchment within a larger basin, which is bounded by large reservoirs and/or reaches in which resource quality objective are not being met. The WDCS may include, but is not restricted to, any of the following water quality variables: (a) *Nutrients* (phosphate, nitrate and ammonium); (b) *Salinity* (total dissolved solids, electrical conductivity, chloride, sodium and sulphate); (c) *pH*; (d) *Heavy Metals* (arsenic, cadmium, chromium, copper, mercury, lead, nickel and zinc); and (e) *Organic material* (Chemical Oxygen Demand).

The WDCS consists of two distinct water use charges, either or both of which may be applied in a specific catchment:

- *Incentive charge* – Charges that provide a disincentive or deterrent to the discharge of waste, based on the use of the resource as a means of disposing of waste.
- *Mitigation charge* – Charges to cover the quantifiable costs of administratively implemented measures for the mitigation of waste discharge related impacts. Mitigation may be more cost effective and acceptable than the incentive charge in reducing the waste load in the catchment. This may be administratively implemented with costs recovered through the mitigation charge.

At time of writing, the WDCS had not yet been introduced, partly due to concerns about the capacity of DWAF to introduce and administer it.

9.5.5 Stream Flow Reduction Activities

While in the past, water charges were only levied for direct water off-stream use activities, the new framework allows any activity which reduces water flows to be charged for. This has been applied in the case of commercial forestry, which is a major user of water. Hydrological research over the past 6 decades in South Africa has shown that commercial plantations of trees such as eucalyptus, pine and wattle in areas that were previously grasslands has resulted in a significant reduction in surface water runoff, reducing water availability to downstream users. This "streamflow reduction" (SFR) due to commercial forests is estimated to be in the order of 1.4 billion m^3 per annum from an area of 1.44 million ha of plantation. Commericial forestry is, therefore, charged for its water use. This approach may be extended to other crops, such as dryland sugarcane or extensive biofuel crops, or to activities such as the construction of small, high evaporation dams in stressed catchments or the failure by land owners to control the spread of water consuming alien vegetation (Box 9.2).

9.6 Conclusion

It is necessary to understand the value-in-use for water resources and the marginal value at which users trade water if one is to gain a true understanding of the role of pricing in water management and try to evaluate the efficacy of pricing policies and allocation decisions.

Fundamental to an effective pricing policy are the pillars of cost recovery, social equity and economic efficiency. DWAF has endeavoured over the past few years to address these aspects in the context of South Africa with a focus on improved quality of life and economic development, through its revisions of the national water pricing

Box 9.2 Kouga example of the working for water programme to improve water flow in rivers

The Kouga site covers an area of 158,678 ha and in the mountain catchment area of the Langkloof valley. The indigenous ground cover is fynobs, but the area is invaded with black wattle and pines. The mean annual precipitation and runoff are 547 and 255 mm per annum, respectively.

Three main dams store water inflows from the Kouga and Krom River mountain catchments which provide more than 80% of the water supply to the Port Elizabeth-Uitenhage metro area and the Langkloof and Patens irrigation schemes. The price of bulk, untreated water from Kouga catchment amounts to US \$0.2 per m^3, which covers the cost of infrastructure not local storage and purification costs. Several new water storage schemes are under consideration or are being developed in the area.

Clearing the Kouga and Drom River catchments of riparian invader plants has been estimated to yield approximately 9 million m^3 water per annum, with no additional cost for water storage, transfer and treatment because such infrastructure already exists. Based on a unit cost of US \$400 per ha for initial clearing and US \$27 per ha per annum for follow up operations, the estimated cost of clearing the 7,000 ha of invaded areas results in an estimated unit value of water of US \$0.03 per m^3, adjusted for inflation to US \$0.05 per m^3 (Hosking and Du Preez 2006).

policies. This includes all tiers of water pricing from infrastructure development to water security for basic needs.

In order to secure the sustainability of water resource infrastructure and the development of new water infrastructure, it is imperative that at least the full cost of water supply be clearly understood and effectively priced and recovered.

The challenge thereafter is to understand the quantum and quality of water required to meet the water needs of the various sectors including but not limited to free basic water, domestic use, industrial use, international obligations, the ecological reserves and agricultural use. Subsidisation requirements of any water user must then be financed from the fiscus or cross-subsidisation systems and not be a loss in income to DWAF or other water supply institutions, which will limit their ability to develop and maintain new water infrastructure.

Such a practice will ensure the long term integrity of water infrastructure and will provide policy makers with an understanding of the economic value of water and an ability to quantify the opportunity cost thereof and its relevant price. This will also enable non-governmental funds to continue to be attracted for development projects that serve large commercial users and allow government's limited finances to be focused on meeting social needs.

References

ANC (1994) The Reconstruction and development programme. African National Congress Retrieved 10 Oct 2007 from http://www.anc.org.za/rdp/rdp.html

Ashton PJ (2002) Water and development: a southern African perspective. In: Slack P, Trottier J (eds) The Linacre lectures – 2002. Oxford University Press, Oxford

Ashton PJ, Seetal A (March 2002). Challenges of water resource management in Africa. Paper presented at the African Renais-Science Conference, Durban

DWA (Department of Water Affairs) (1970) Report of the Commission of Enquiry into water matters. R.P. 34/1970. Republic of South Africa, Pretoria

DWA (Department of Water Affairs) (1984) White paper on the policy on water tariffs and related matters. WP-84, ISDN 062085855, Pretoria

DWAF (Department of Water Affairs & Forestry) (1997) Overview of the water resources availability and utilisation in South Africa. Department of Water Affairs and Forestry, Pretoria

DWAF (Department of Water Affairs & Forestry) (1999) Resource directed measures for protection of water resources. Volume 3: River ecosystems version 1.0. Department of Water Affairs & Forestry, Pretoria

DWAF (Department of Water Affairs & Forestry) (2002) Water resource protection policy implementation, methodology for the determination of the preliminary ecological reserve for estuaries, Version 2. Department of Water Affairs and Forestry, Pretoria, South Africa

DWAF (Department of Water Affairs and Forestry) (2003) Strategic framework for water services. Department of Water Affairs and Forestry, Pretoria, South Africa

DWAF (Department of Water Affairs and Forestry) (2004) National Water Resources, Strategy Department of Water Affairs and Forestry, Pretoria, South Africa

DWAF (Department of Water Affairs and Forestry) (2004) Introduction to the Orange River Basin. [Electronic version]. Pretoria, Department of Water Affairs and Forestry. Retrieved 7 June 2008 from http//www.dwaf.gov.za

DWAF (Department of Water Affairs and Forestry) (2007) Establishment of a pricing strategy for water use charges Government Gazette No. 25697 16 March 2007

Hosking SG, Du Preez M (2006) Valuing water gains in the Eastern Cape's working for water programme. Department of Economics and Economic History, University of Port Elizabeth

International Monetary Fund (2007) World economic outlook database

King NA (2002) Valuing a city's water: the case of Tshwane. MCom Thesis, University of Pretoria, Pretoria

Muller M (2007) Parish pump politics: the politics of water supply in South Africa. Prog Dev Stud 7:1, 2007

Muller M (2008) Free basic water – a sustainable instrument for a sustainable future in South Africa. Environ Urban 20(1):67, 87

Priscoli DJ (Aug 1998) Water and civilization: conflict, cooperation and the roots of a new eco-realism. In: Proceedings of the eighth Stockholm world water symposium, Stockholm, Sweden

RSA (1996) The constitution of the republic of South Africa. Government Printer, Pretoria, Act No. 108 of 1996

RSA (1997) White paper on a national water policy for South Africa. Pretoria, Government Printer

RSA (1998) National water act. Government gazette, vol 398. Office of the President, Cape Town, 26 August 1998

RSA (2000) The municipal systems act, vol 398. Republic of South Africa, Pretoria, Act. No. 32 of 2000

Smakhtin V, Ashton PJ, Batchelor A, Meyer R, Maree JP, Murray M, Barta B, Bauer N, Terblanche D, Olivier J (2001) Unconventional water supply options in South Africa: possible solutions or intractable problems? Water Int 26(3):314–34

Umgeni Water (2005) Part one policy statement – business plan. Presentation to DWAF, Pretoria

Chapter 10
Mainstreaming Gender in Water Management in South Africa

Barbara van Koppen, Barbara Schreiner, and Eiman Karar

Abstract Gender mainstreaming figures high in the post-1994 policies and laws in South Africa in general, and water policies in particular. This chapter analyses the implementation of these policies in two domains: within DWAF as a gender-sensitive workplace with sound gender training of its staff, and externally in the performance of DWAF in implementing its mandate for the benefit of all its citizens, in particular poor black women. In this task, gender concerns were effectively mainstreamed as part of the general efforts to democratise water management, especially in the creation of new equitable institutions such as Catchment Management Agencies and in public participation processes. Changes appeared more difficult in existing male-dominated institutions, though. With regard to the core issue of improving women's access to water, the water services efforts implicitly benefitted women in particular. In contrast, women's access to water for small-scale productive uses has deteriorated.

Keywords Affirmative action • Gender • Multiple water uses • Water policy and law

10.1 Gender Mainstreaming in South Africa

Within less than a decade, water resources management in South Africa has been transformed from a male-dominated domain, serving the interests of a white minority, to a dynamic and creative process which encompasses forms of gender mainstreaming

B. van Koppen (✉)
International Water Management Institute, Pretoria, South Africa
e-mail: b.vankoppen@cgiar.org

B. Schreiner
Pegasys Strategy and Development, Pretoria, South Africa
e-mail: barbara@pegasys.co.za

E. Karar
Water Research Commission, Pretoria, South Africa
e-mail: eimank@wrc.co.za

B. Schreiner and R. Hassan (eds.), *Transforming Water Management in South Africa: Designing and Implementing a New Policy Framework*, Global Issues in Water Policy 2, DOI 10.1007/978-90-481-9367-7_10, © Springer Science+Business Media B.V. 2011

that have gained global recognition. This chapter analyses the process of gender mainstreaming in water resources management since 1994. 'Gender' has many dimensions. It refers to the political, socio-economic, cultural and psychological attributes that societies ascribe to humans of different biological sexes. Gender intersects many other axes of social stratification including caste, class, race, age, and marital status, as well as physical and mental ability (Lahiri-Dutt 2006). Gender mainstreaming brings gender relations to the centre stage of societal change. This chapter takes a political economic perspective of gender and its interconnectedness with race, and understands gender mainstreaming against the background of broader social change.

In South Africa, gender mainstreaming in water management was part of a greater gender mainstreaming process in society, as a whole, both before and after the first democratic elections. Prior to 1994, women from all classes played key roles in the anti-apartheid movement. In the early 1990s when a new Constitution, other laws, policies, and state structures were designed, the same gender activists, supported by progressive men, seized the opportunity to mainstream gender, like race, wealth, status, sexual preference, and disability, into the new institutional landscape. The Constitution (RSA 1996) is one of the most progressive in the world for enshrining gender equality. It also recognises that non-discriminatory policy and legislation alone are not sufficient, but that affirmative action is required to end past discrimination (Schreiner and van Koppen 2003). All laws and programs should comply with the Constitutional imperatives of gender equity, such as labour laws as well as land tenure and land reform policies, supported by a myriad of nation-wide campaigns against the abuse of women and children, or encouraging girls to take up professional careers. Gender also features high in the new provincial and local government institutions, and the new policies and laws of the line departments, including the Department of Water Affairs and Forestry, are the focus of this chapter.

While South Africa's gender mainstreaming is articulated in policies and laws, a series of institutional arrangements have been put in place to ensure their implementation and enforcement (Office of the Status of Women 1999). The critical issue of enforcement is institutionalised through gender agencies that play both an advisory and a 'watchdog' role – albeit with limited resources. At national level the latter role is taken up, among others, by the Office of the Status of Women in the Presidency, an independent Commission on Gender Equality, and a Parliamentary Committee on the Status of Women. Many parliamentary portfolio committees, including the one for water affairs, also scrutinise the implementation of gender policies.

The government's gender commitment is reflected in Cabinet: with 42% women in the Cabinet, South Africa is second only in the world to Sweden (Lowe Morna 2007). South Africa is also the ready signatory of international conventions, like the Convention on the Elimination of all Forms of Discrimination against Women (CEDAW) of the Office of the UN High Commission for Human Rights, which was ratified in 1995. Regionally, South Africa is active in moving the already ratified 'nice to do' SADC Declaration on Gender and Development into a more binding 'have to do' SADC Protocol (Lowe Morna 2007).

This context of transition largely shaped and continues to shape gender mainstreaming in the Department of Water Affairs and Forestry (DWAF). When the Department set out to transform the state apparatus, originally designed to serve the minority and to marginalise the majority, into a unitary state serving all, gender mainstreaming also became central in the transformation of water resources management. The National Water Act (1998) reflects this commitment in its main goals, which are; 'promoting equitable access to water' and 'redressing the results of past racial and gender discrimination', and [...] 'for achieving this purpose, to establish suitable institutions and to ensure that they have appropriate community, racial and gender representation.' (NWA, Section 1.2). DWAF's two male and two female gender-activist ministers played a key role in implementing the government's commitment to gender equity in DWAF, both at the workplace and in its delivery mandates. In 2005, a Gender Directorate was established within DWAF to drive and monitor gender-mainstreaming activities.

In the further operationalisation of gender mainstreaming in DWAF, two components are distinguished. As elaborated on in Section 10.2, the one is in-house and ensures gender mainstreaming in DWAF as the workplace, in addition to ensuring that all staff are trained in gender issues. The other component is outward-looking and concerns gender mainstreaming in DWAF's developmental and regulatory services to the nation, in particular, to poor black women who bore the brunt of the apartheid political economy and constitute the majority of the poor today. Section 10.3 takes the legacy of water resource management from the past as a starting point and illustrates how this was opened up, from the top-down, to reach a more race- and gender-balanced clientele. Section 10.4 addresses perhaps the greatest challenge which is, for the first time in history, actually improving access to and control over water by Historically Disadvantaged Individuals, both for domestic and productive uses.

To conclude, the context of the transformation in water resources management in South Africa was unique and came with great creativity and energy in the drive to mainstream gender, like race and equity in wealth. Yet the challenges on the table are the same challenges that gender activists, governments, civic society, and the international water and development community face elsewhere. The concluding Section 10.5, summarises lessons learnt that may also be relevant beyond South Africa.

10.2 Gender and Race Mainstreaming Within DWAF

National policy and legislation required DWAF to transform into a gender-sensitive workplace. A great deal of work was done in this regard, including affirmative action, to drive gender equity in staffing. DWAF's Gender Policy (DWAF undated) stipulates a number of gender provisions, such as the requirement for affirmative action in the Department to 'ensure that women in the Department are given the support, training and capacity-building to fulfil their work requirements'. It also states that there should be affirmative action to ensure gender equality in the staff employed in the department, and that attention should be given to the requirements of female staff who are still the primary care-givers within the family (Schreiner and Van Koppen 2003).

As shown in Table 10.1, by 2005 affirmative action in gender and race staffing had led to major changes in staff composition within DWAF, compared to an almost exclusively white male composition a decade earlier. Many of the latter retired or joined engineering and consultancy firms – often to offer their services to DWAF in this way.

In order to better recognise women's achievements within and outside DWAF, prestigious annual Women and Water Awards were initiated in 2002. This initiative is driven by DWAF, the Water Research Commission, the electricity parastatal, Eskom, and other sponsors.

Gender mainstreaming was also included in job descriptions and performance agreements. Old and new, male and female DWAF officials were required to include gender- and race-considerations in their jobs according to DWAF's new mandate. In 2005/2006, DWAF organised in-house gender training for its senior staff in water resources policy and regulation. This training took managers' own function description as the starting point for gender mainstreaming.

All DWAF staff were encouraged to innovate in the field of gender by rewarding pro-active gender-equitable service delivery through annual gender prizes, handed out by the Minister. Innovations covered the entire range of the Department's activities. In 2005, for example, they encompassed training in women's assertiveness in the Eastern Cape, joining national initiatives to promote girls' career planning in Northern Cape, innovative targeting approaches in national forestry programs to empower women for bee-keeping and honey production – which was traditionally a male domain, promoting women as the better-performing contractors of labourers in Working for Water programs for eradicating alien, water-scavenging trees and shrubs in Limpopo Province, and the promotion of women in new businesses for the biological composting of mine debris through worms in North West Province. The Water Research Commission initiated various studies to evaluate gender-sensitive service delivery and feed the results back to DWAF. They highlight some progress, but also major challenges that are still to be addressed.

10.3 Gender in the Democratisation of Water Resources Management

A gender-sensitive workplace and gender expertise among DWAF staff contribute to the other gender domain discussed in this chapter: in implementing DWAF's new mandate externally to its female and male clientele across the nation, DWAF's gender performance was embedded in the broader changes. As discussed in Chapter 1, at the dawn of democracy, DWAF inherited a system of water resources management that primarily served the white economy. Until about the 1970s, the state had

Table 10.1 Race and gender composition of DWAF by 2005

Race	Gender	%					Total		
		Political office bearer (Minister)	Top management	Professional qualified/ specialist	Skilled technical workers	Lower level/ production	Number	%	Total population in South Africa in 1996
African	Female	1.0	6.4	14.4	17.4	30.5	*4,639*	28.4	
	Male	0.0	20.8	31.4	43.1	63.3	*9,753*	59.7	
	Total	1.0	27.2	45.8	60.6	93.8	*14,392*	88.1	76.0
Coloured	Female	0.0	0.8	1.5	1.2	0.5	*95*	0.6	
	Male	0.0	2.4	1.6	2.3	2.9	*450*	0.8	
	Total	0.0	3.2	3.1	3.5	3.3	*545*	3.4	8.0
Asian	Female	0.0	0.8	1.5	1.1	0.1	*48*	0.3	
	Male	0.0	4.0	1.5	1.0	0.1	*47*	0.3	
	Total	0.0	4.8	3.0	2.1	0.2	*95*	0.6	3.0
Black	*Female*	*100.0*	*8.0*	*17.4*	*19.8*	*31.1*	*4,782*	*29.3*	
	Male	*0.0*	*27.2*	*34.5*	*46.4*	*66.3*	*10,250*	*62.8*	
	Total	**100.0**	**35.2**	**51.9**	**66.2**	**97.4**	***15,032***	**92.1**	**87.0**
White	*Female*	*0.0*	*4.8*	*9.8*	*12.0*	*1.5*	*457*	*2.8*	
	Male	*0.0*	*60.0*	*38.3*	*21.8*	*1.1*	*843*	*5.2*	
	Total	**1.0**	**64.8**	**48.1**	**33.8**	**2.6**	***1,300***	**8.0**	**13.0**
All	**Female**	**100.0**	**12.8**	**27.2**	**31.8**	**32.6**	***5,239***	**32.0**	**52.0**
	Male	**0.0**	**87.2**	**72.8**	**68.2**	**67.4**	***11,093***	**68.0**	**48.0**
	Total	100.0	100.0	100.0	100.0	100.0	*16,332*	100.0	100.0

Source: Prepared with information from DWAF June 2005, Garduno and Hinsch 2005.

mainly supported dam construction and white irrigated farming throughout the country, while mining houses, local governments and other white constituencies self-provided water supplies at lower scales. With the agricultural transition among whites and further urbanisation and industrialisation, in particular in Gauteng, South Africa's economic hub, the role of the then Department of Water Affairs (DWA) shifted. While the best sites for dam construction had been taken, the needs for high-security and year-round water supplies to the elevated plateau of Gauteng grew rapidly. DWA engaged in the long-term planning and partial operation and management of increasingly-extended and complex inter-basin bulk water supplies for a white clientele. By then, this clientele had become wealthy enough to pay for even such costly services. Notions of water as an economic good, user pays, allocation of water according to market principles, etcetera, accompanied this shift. In this period, DWA also recognised the need for more regulation, in particular with regard to mining and urban pollution, demand management, and 'the environment' (which was defined as nature reserves and conservation parks) (DWA 1970; Backeberg 1994; Van Koppen 2007).

However, the large majority of Historically Disadvantaged Individuals had been systematically excluded from access to water and the benefits from water. Hardly any black man, let alone a black woman had a water entitlement in his or her name. Access to water in townships and former homelands was, at best, mediocre. Benefits of water through employment hardly trickled down because wages were extremely low – and lowest for black women. Moreover, from the 1970s onwards, high levels of unemployment emerged in this lop-sided capital-intensive high-tech minority economy.

After 1994, DWAF deployed major efforts to open up and democratise this 'white water economy' while keeping its centralised top-down mandate. In the various initiatives to this end, gender issues received considerably more attention than ever before. This is well illustrated in the most important innovation in basin water resource planning and management, as prescribed in the National Water Act (1998) – 'extensive public participation in the formation of new Catchment Management Agencies with Catchment Management Strategies that must align with the National Water Resource Strategy. In order to ensure gender and racial representation in the Governing Board of a Catchment Management Agency, the Act gives the Minister the power to appoint the members to the Governing Board as nominated by an Advisory Committee. Also after that, s/he is entitled to 'appoint additional members selected by the Minister in order to [...]

(b). achieve sufficient gender representation
(c). achieve sufficient demographic representation [...]
(e). achieve representation of disadvantaged persons or communities which have been prejudiced by past racial and gender discrimination in relation to access to water.' (NWA, Chapter 7 Section 81–10)

DWAF initiated processes of public participation, both at national level for the first National Water Resources Strategy, and, with varying intensity, at catchment level to consider or prepare proposals for Catchment Management Agencies.

Through continuous cycles of information and outreach, invitation, facilitation, and accounting for inputs made, water resources management changed from an uncontested right and responsibility of a minority, to a democratised process in which the legitimacy of any planned activity is scrutinised by a much larger group of whites *and* blacks than ever before. Although still a minority, women started participating as well. For example, in the public participation processes of the National Water Resource Strategy in 2002–2004, 57% of the participants were whites, and 15% of the whites were women. Among the 43% of blacks who participated, one quarter was women (Schreiner and Van Koppen 2003). In the first CMA, the Inkomati, the Board consists predominantly of HDIs, and almost 50% of the 44 staff are women, although the percentage is much lower at senior management level (ICMA 2007).

It would appear that it was easier to comply with the quota of women in brand-new institutions, like CMAs, than in existing institutions. This became clear in the transformation of Water User Associations (WUA). The National Water Act offers the possibility to set up a WUA voluntarily, for local water management of mutual interest. The management committees of WUAs are expected to be representative in relation to race and gender. Former Irrigation Boards were obliged to transform into such WUAs. The quota for blacks and women in the management committees was set at 50%. However, implementation has been disappointing, and difficult. Out of the 59 registered Water User Associations in 2006, only a few were made up of black emerging farmers and other HDI groups. In a study of half of the new WUAs, only 15% of the WUAs had more than 30% representation of women. Percentages are lowest in transformed Irrigation Boards. Most have less than 10% women (Mjoli and Njiro 2007). Faysse (2004) found the same in seven transformed Irrigation Boards. Five have not a single woman on their management committees. In the other two, the figures are only 10% and 12%. This reflects the continuing legacy of traditional male dominance in white irrigation and the continued ownership of land and access to water by white men in the commercial farming sector.

10.4 Gender in Improved Access to Water

10.4.1 Domestic Water Services

Perhaps an even greater challenge for gender mainstreaming than introducing participatory processes, was redressing the historical and systematic exclusion of black South Africans, particularly black women, from access to water. The importance for women and girls of improved water services for domestic uses does not need any further explanation. Hence, the new government's immediate priority to improve access to water services was key to the alleviation of poor women's drudgery of water-fetching. As described in detail in Chapter 3, the White Paper on Water Services articulated how the backlogs in domestic water services delivery were to be addressed as a priority. DWAF put in place a major programme for the delivery of water services to unserved rural communities. By the mid-1990s, new provinces with their provincial governments had been established, but the complex process of demarcating the boundaries of 284 new municipalities, and electing new councillors and mayors was yet to be finalised, as happened in 2000. The formal take-over by municipalities of the DWAF-driven Community Water Supply and Sanitation programme took place in 2006. This hand-over and the reformulation of DWAF's role towards capacity building and regulation were described in the Strategic Framework for Water Services of 2003.

Achievements have been substantial. According to DWAF's figures, access to basic water supply services improved from 59% of the population in 1994 to 83% in 2005, so that South Africa achieved the Millennium Development Goal of halving the proportion of people without access to water ahead of schedule (DWAF 2006c).

Gender mainstreaming kept featuring high in the national 'Implementation Strategy and Action Plan for 2006–2010'. This plan includes; policy formulation, institutional support, programme management, resource mobilisation, communication management, internal transformation, economic transformation, partnerships and networking, monitoring, evaluation and reporting, capacity building, advocacy and awareness raising, as well as research and development. It also raised the quota for women's representation in domestic water committees from one-third to half.

10.4.2 Improving Water for Productive Uses and Multiple Uses

Improving the access to water for productive uses by the 'have-nots' appeared more complicated. Yet, the importance of improving access to water for poverty eradication, and especially women's empowerment, cannot be overestimated. In rural and peri-urban areas, black women constitute the majority (up to 90%) of farmers for gardening, cropping, tree growing and small enterprises, and accordingly, have stronger water needs (Makhura and Ngqaleni 1996; Stimie and Chancellor 1999; Van Koppen et al 2006). While DWAF provided a temporary solution for domestic water services delivery, no such cushion existed for irrigation schemes in the former homelands. Here, the sudden withdrawal by the Department of Agriculture of even the mediocre support by the apartheid regime led to the deterioration, if not collapse, of most schemes (Shah et al 2002).

The revitalisation or new construction of irrigation schemes is only slowly getting back to speed. At the national level, DWAF launched a subsidy scheme in 2004 to support resource-poor farmers (DWAF 2007). This enabled, among others, the roll-out of rainwater harvesting tanks in 26 villages in four provinces. DWAF also engaged with many smaller-scale civil society and donor initiatives with important gender components. Links were maintained with NGOs, like the Water for Food Movement, around homestead-based water harvesting and household food security by the poorest of the poor. DWAF started collaborating with donors like the SADC/Danida-supported Integrated Water Resource Management program, which also articulates gender issues. The Water Research Commission also initiated new studies on, for example, rainwater harvesting, multiple uses of piped water supplies among poor households, gender in land tenure and productivity.

A major challenge for government in moving from spontaneous small-scale 'let all flowers bloom' initiatives, to a more comprehensive program is that its delivery capacity on the ground is even weaker than the limited water services delivery capacity. Water infrastructure for productive uses requires inter-departmental co-ordination, aligned around Integrated Development Plans at municipal level, Provincial Strategies for Growth and Development and Provincial Water Services Plans. The horizontal and vertical alignment of these new government institutions, as promoted, among others, by the Governmental Relations Framework Act of 2005, may take decades to crystallise around the effective delivery of water for productive purposes. At national and provincial level, DWAF reached out to forge such relationships with the National Department of Agriculture and Land Affairs and, more systematically, with provincial and local government. In 2005 and 2006, DWAF organised Provincial Water Summits and a synthesising National Water Summit. These new forums allowed a wide range of stakeholders, including emerging black farmers, male and female, to express their water needs, and in a relatively representative and bottom-up way.

A major advantage in promoting service delivery for productive water uses through provincial and local government is that it can be grafted upon already-existing relationships and processes for domestic water services delivery. In this synergy, it is becoming clear that the separation between domestic and productive water uses is not meaningful in rural households and communities where water is used, simultaneously, for domestic uses *and* small-scale gardening, cropping, livestock, trees, and small businesses. Multiple water uses occur especially at and around the homesteads (Perez de Mendiguren Castresana 2004). At and around homesteads, women benefit even more than men from access to water, because women are responsible for domestic water provision and tend to rely more on homestead production than men. As caretakers of the sick, including AIDS victims, women also benefit from ample water provision near the house to meet the greater water demands of the sick and to enable the weaker to grow food. DWAF drafted guidelines to promote such multiple-use water services ('MUSs') nation-wide. A methodology to that end has been piloted in Bushbuckridge Local Municipality. Based on a holistic participatory assessment of all water resources, technologies, and multiple uses, and after the identification and prioritisation of multiple water needs, proposed actions are incorporated in the municipality's Integrated Development Plan (Maluleke et al 2005). The call for better integration of water services and water resources management is taking place at all scales, up to national level policy making and legislation.

A more structural obstacle to improve rural women's access to water for productive uses relates to the broader political and economic context of the new South Africa. Gender mainstreaming in water management and society as a whole requires, above all, establishing a gender- and race-equitable pattern of economic growth and poverty eradication. Under apartheid, the majority of black women were relegated to underpaid wage labourers and low-productivity subsistence farmers on land belonging to their male kin in congested homelands, without state support, and facing strong market competition from heavily subsidised white commercial farmers and food retailers. Redressing such an extractive, high-tech and capital-intensive pattern of economic growth into broad-based economic growth is a tall order. It raises questions of how water management can contribute to, for example, gender-equitable employment in labour-intensive, rather than capital-intensive, production in both formal and informal economies. To achieve economic equity, the issue of agrarian reform must be addressed so that land and water reform and other services can collaborate towards an agrarian structure which is considerably more equitable along wealth, gender, and race lines. Agrarian reform is most relevant for black women, given their existing strong roles in cropping, and the high level of poverty in female-headed households. As increasingly acknowledged in the land tenure debates (World Bank 2008), women are as efficient producers as men, provided they have equal access to land, inputs, markets, technologies, and water (Van Koppen 2002). Under growing competition for water, water development and allocation, policies for broad-based gender-equitable economic growth in rural areas become even more pivotal.

10.5 Conclusions

The following conclusions can be derived from the unprecedented experimentation, momentum, and commitment to mainstreaming gender in South Africa's watershed political and institutional changes:

- Gender equity is not given but taken by persistent struggle and strategic insight. This started in the anti-apartheid struggle and continued during the formulation of new policies, laws, programs and structures.
- Gender-equitable policies and laws are indispensable but toothless without their continued enforcement and monitoring by the country's highest political structures.
- Political support aided gender mainstreaming within DWAF, but this also required pro-active promotion by senior management.
- Setting quotas, affirmative action and measuring recruitment programs' performance for more gender- and race-balanced staffing, works.
- Gender expertise, within and outside the water department is needed for training of male and female staff, gender disaggregated data collection, monitoring and scientific evaluation for feedback.
- Top-down commitment and resource-allocation can achieve gender equity in newly established organisations. However, without strong leadership, transforming existing male-dominated institutions into more gender-equitable institutions is slow.
- Promoting multiple water uses at and around the homestead tends to benefit women more than men.

- The prominent role of women smallholders in South Africa indicates a powerful entry point for agrarian change and broad-based economic growth steered by their enhanced access to and control over land and water.

Acknowledgement This chapter was prepared with the support of PN17 "Integrated water resources management for improved rural livelihoods", a project of the CGIAR Challenge Program on Water and Food.

References

Backeberg GR (1994) Die politieke ekonomie van besproeiingsbeleid in Suid-Afrika. Voregle ter vervulling van 'n deel van die vereistes vir die graad Philosophiae Doctor. Departement Landbou-ekonomie. Voorligting en Lndelijke Ontwikkeling. Fakulteit Biologiese en Landbouwetenskappe. Universiteit van Pretoria, Pretoria

DWA (Department of Water Affairs) (1970) Report of the commission of enquiry into water matters. Republic of South Africa, Pretoria, R.P. 34/1970

DWAF (Department of Water Affairs & Forestry) (2006) National overview of the outcomes and actions. Department of Water Affairs and Forestry, Pretoria, Consolidated report on Provincial Water Summits. May 2006

DWAF (Department of Water Affairs and Forestry) (2007) Regulations on financial assistance to resource poor farmers. Government Notice. R.1036 National Water Act (3611 998): no. 3 30427

Faysse N (2004) An assessment of small-scale users' inclusion in large-scale water users associations of South Africa. IWMI Research Report 84. IWMI, Colombo

ICMA (2007) Annual Report 2006/7. Inkomati Catchment Management Agency, Nelspruit

Lahiri-Dutt K (ed) (2006) Fluid bonds: views on gender and water. Stree, Calcutta

Lowe Morna C (2007) Genderlinks http://www.genderlinks.org.za

Makhura T, Ngqaleni MT (1996) An analysis of women's status in agricultural development in the Northern Province. In: Lipton M, Ellis F, Lipton M (eds) Land, labour and livelihoods in rural South Africa. Chapter 13, vol 2. Indicator Press, Durban

Maluleke T, Cousins T, Smits S (2005) *Se*curing water to enhance local livelihoods (SWELL): community-based planning of multiple uses of water in partnership with service providers: a case study on its application in Bushbuckridge. South Africa. Retrieved 19 May 2008 from http://www.musproject.net/page/533

Mjoli N, Njiro E (2007) Assessment of gender equity within water user associations in South Africa. Status Review Report. K8/754. Water Research Commission, Pretoria

Office of the Status of Women (1999) National policy framework for women's empowerment and gender equality. The Presidency of the Republic of South Africa, Pretoria

Perez de Mendiguren Castresana JC (2004) Productive uses of water at the household level: evidence from Bushbuckridge, South Africa. In: Moriarty, Patrick, John Butterworth, Barbara van Koppen (eds) Beyond Domestic. Case studies on poverty and productive uses of water at the household level. IRC Technical Papers Series 41, IRC, NRI, and IWMI, Delft

RSA (1996) The Constitution of the republic of South Africa (Act No. 108 of 1996). Government Printer, Pretoria

Schreiner B, van Koppen B (2003) Policy and law for addressing poverty, race, and gender in the water sector: the Case of South Africa. Water Pol Official J World Water Council 5(5/6):489–501

Shah T, van Koppen B, Merrey D, de Lange M, Samad M (2002) Institutional alternatives in African smallholder irrigation: lessons from international experience with irrigation management transfer. IWMI Research Report No 60. IWMI, Colombo

Stimie C, Chancellor F (1999) Smallholder irrigation – South African women's part in rehabilitation and design of irrigation projects. GRID Issue 13, February 1999

Van Koppen B (2002) A gender performance indicator for irrigation: concepts, tools, and applications. Research Report 59. International Water Management Institute, Colombo

Van Koppen B (2007) The basin development trajectory of the Olifants basin in South Africa before 1994. Unpublished. Water Research Commission, Pretoria, South Africa

Van Koppen B, Khumbane T, de Lange M, Mohapi N (2006) Gender and Agricultural Productivity: Implications for the Revitalization of Smallholder Irrigation Schemes Program in Sekhukhune District, South Africa. In: Lahiri-Dutt (ed) Fluid bonds: views on gender and water. Stree, Calcutta

World Bank (2008) World Development Report 2008. Washington DC: World Bank. Retrieved from http://go.worldbank.org/LBJZD6HWZ0

Chapter 11
The Role of Information Systems Management in the Management of Water

Mark Dent

Abstract The primary aim of water information systems management is to serve the social processes which give the information systems context and meaning. The social processes related to water in South Africa have changed substantially in the past 25 years and particularly since 1994. This chapter will consider the past, present and future roles of information systems management in the context of these changes. The central issues of access, affordability, and balancing power relations with respect to information in an increasingly complex web of integrated water and social systems are key themes in this chapter.

The subject of water has many organizational and disciplinary facets. It is not the intention of this chapter to deal with the role of information in each of these in detail. The approach has been to explain and explore the key strategic elements of the role of information systems management in the management of water in South Africa, particularly in relation to water resources management, although many of the messages which follow are generic and apply to both the water services and water resources realms. The decision to distil out the generic messages has been a conscious one that is consistent with approaches to deepen understanding of management issues in situations where a focus on specific content may blind the reader to key management issues. It is commonly stated that 'information is power'. This is particularly true in the complex historical, geographical and water setting of South Africa. Issues of access, affordability, complexity, interpretation and balancing knowledge power relations in the dynamic, democratic interplay amongst role players, with respect to water related information, are of central importance.

Keywords Systems • Information • Modeling • Power • Knowledge • Interests • Integration

M. Dent (✉)
Centre for Environment, Agriculture & Development (CEAD),
University of KwaZulu Natal, KwaZulu Natal, Durban, South Africa
e-mail: dent@ukzn.ac.za

B. Schreiner and R. Hassan (eds.), *Transforming Water Management in South Africa: Designing and Implementing a New Policy Framework*, Global Issues in Water Policy 2, DOI 10.1007/978-90-481-9367-7_11, © Springer Science+Business Media B.V. 2011

11.1 Introduction

It is important to emphasise the range and number of South African stakeholders, from every conceivable avenue of human endeavour, that have a deep interest in water. A natural outcome of this widespread interest is that water-related data collection and information generation is carried out by many state and private organisations throughout the country. So, whilst the Department of Water Affairs and Forestry (DWAF) has the responsibility for managing South Africa's water resources, much of the information that they collate and store is in fact collected by other organisations. These organisations are primary custodians of the data and information, whilst DWAF is the secondary custodian. Thus, most of the data and information is held in relationship between two and often many more organisations, bringing into play a variety of management issues that would not be present if DWAF was the primary collector and custodian of the data. A further complication is introduced when information is derived by combining data and or information from several organisations into one information product – who owns the hybrid product, and who trusts the hybrid product? Who pays for the data collection and or information generation? In many cases, the South African tax payer pays and the data and information is, strictly-speaking, in the public domain. Relationships at all levels are, therefore, a key element of water information management, making it a multi-faceted, complex and dynamic subject, especially in the South African context.

11.2 Overall Context and Historical Perspective

The primary aim of water information systems management is to serve the social processes which give the information systems context and meaning. The social processes related to water in South Africa have changed substantially over the past 25 years and particularly since 1994. The preceding chapters of this book address these changes in detail and, as such, provide a vital context for this chapter. The central issues of access, affordability, and balancing power relations with respect to information in an increasingly complex web of integrated water and social systems are key themes in this chapter. This chapter will explore the key strategic elements of the role of information systems management in the management of water in South Africa.

Local, national and international political change has shaped the way societies manage water. Information technology is in the forefront of the strategic forces shaping water information management. However, the direction of change induced by the information technology revolution is not predestined as may be presumed. Quadir et al. (1999), writing for the Global Water Partnership which was developing a World Water Vision for 2025, made the following statement which is equally true for South Africa:

> The impacts of information technology on the water sector are not inherent in the technology but largely depend on the way society chooses to use the technology. The new technology does offer unprecedented possibilities to change knowledge relationships

which impact on power relationships and consequently on organisations and society at large (Quadir et al. 1999).

In Fig. 11.1 below, Turton (1999) depicts some of the changes that influence societal choices in water management. Reflection on this diagram, in the light of the statement by Quadir et al. (1999), reveals some of the complexity of the realm of human endeavour in which water-related information must be managed.

In the diagram below, South Africa falls in the Demand and Adaptive phases. In addition to the usual discursive elites, there are also the new democratic processes of stakeholder interaction required by the National Water Act (NWA). Understanding the emerging role of information and knowledge power in this process is the key to understanding the intended democratic process of water management in South Africa.

In the light of the statement by Quadir et al. (1999) and the trends illustrated by Turton (1999), it is ironic that information management is one of the elements in the integrated water management realm that has, so far, changed the least in South Africa, whilst the technology supporting it has changed so much. The potential for substantial change in the way the water sector stakeholders manage information is now high.

Possibly the most powerful strategic force acting on the water realm, and hence on water information systems management, is the imperative for co-operative government and co-management. Chapter 3 of the Constitution requires such co-operation. Implied, and at times explicit, in water policy and legislation is the requirement that water information management systems serve the socio-political processes in an integrated, affordable, transparent and accessible manner.

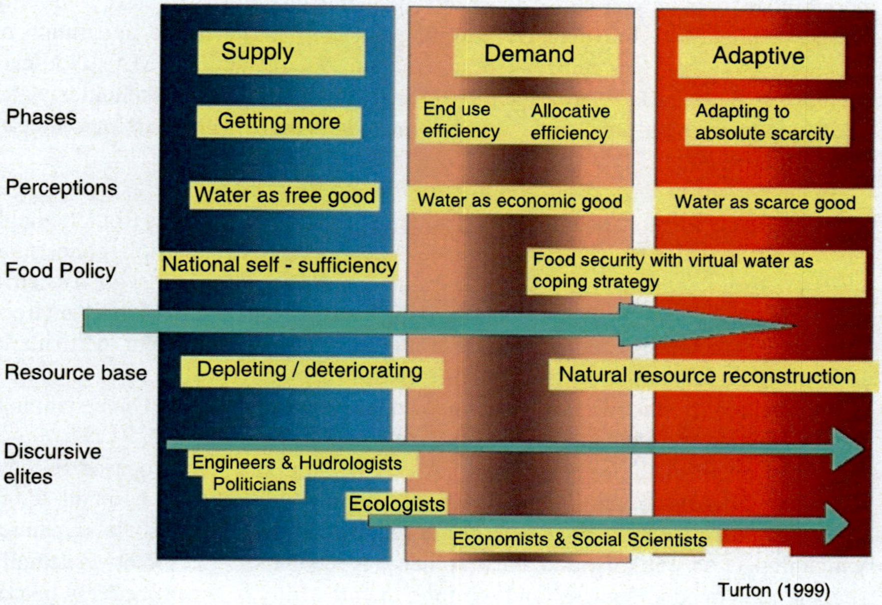

Fig. 11.1 Major changes in the field of water management (after Turton 1999)

At the same time, the world-wide increase in consciousness of environmental matters, including water, has resulted in the private sector being subjected to more pressure, mainly through conventions, protocols and marketing imperatives, to take water-related matters more seriously. This has led to them employing more water-skilled personnel, and commissioning more water-related work from the consulting sector. This trend was reported as early as 1998 by the DWAF/UNESCO/WMO (1998) mission to assess the education and training needs of the water resources management sector of South Africa. Water-related knowledge power has continued to shift from the public to the private sector. The exceptionally high percentage of outsourcing to consultants is evidence of this shift in water-related knowledge. The consequences for the management of water information systems are important. There are also policy, legislative and regulatory frameworks which have spawned new multi-stakeholder institutional forms, such as Catchment Management Agencies (CMAs), Water User Associations (WUAs) and Catchment Management Forums (CMFs), in the quest to democratise water management.

Effective information systems are key to effective working together in water management. Other industries, notably information technology and telecommunications, aerospace and the related travel industry, and the financial services and banking industry with ATMs, credit cards and on-line payment systems, provide excellent lessons in this regard. All these information systems allow a plethora of independent, yet interdependent, organisations to operate together seamlessly.

The NWA enables the decentralisation of water resources management, giving more power to stakeholders. These new institutional arrangements have created possibilities for economies of scale, critical mass and institutional memory-building that have previously been elusive. Exciting possibilities lie in the integration of previously-fragmented water-related information systems within CMAs. Not least of these is the possibility for such systems to serve the multi-stakeholder water allocation bargaining process, in which knowledge equity will be as important as representative equity.

In the field of water, the vast majority of the numbers that are useful for management are, in fact, generated information and not primary observations from the field. South African policy and legislation mandate that management decisions concerning water are made, primarily, in a socio-scientific context. In other words, good science must serve into a healthy social process around decision-making. In such an environment the distinction between data and information is vitally important, particularly in terms of the assumptions made in creating information from data.

Framed in the constructs of Relational Connection and Systems Understanding, the progression from data to wisdom is shown in Fig. 11.2 below.

Whilst much of the effort in South Africa has, to date, been focused on data collection, it is information, knowledge and, finally, wisdom which social actors seek. The possession of data is woefully inadequate to manage the social dynamics which impact on water use and abuse. When it is recognised that society is actually seeking integrated knowledge and wisdom, then the folly of introspectively hoarding, and not sharing, data becomes apparent.

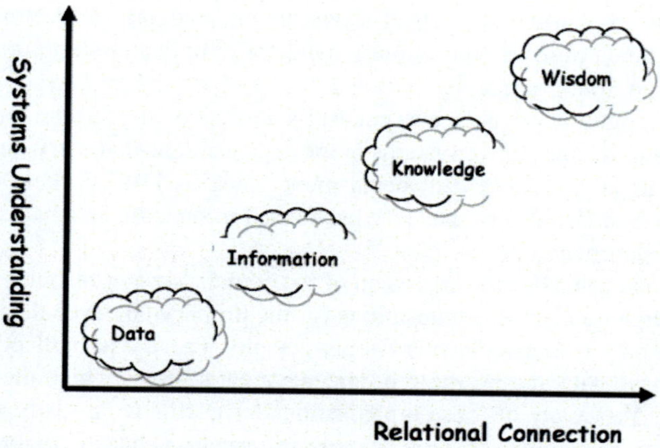

Fig. 11.2 The progression from data to wisdom, framed in the construct of Relational Connection and Systems Understanding

However, Senge et al. (1995) stress that more information is not necessarily better for decision-making, and that information is first interpreted through the "lenses of the human mind", before its value and relevance to the debate is understood and verbalised. They point out that the notion that all that managers need for better decision-making is more information, misses the critical aspect of information interpretation. According to Sterman (1989) research results directly contradict the assumption that all that managers need for better decision-making, is more information. He reported that many studies have found that even knowledgeable and experienced decision-makers often filter their information through non-systemic mental models, construing symptoms as causes and reacting in ways that make problems worse rather than better. Senge et al. concede that increased access to information may be a step in the direction of enhancing learning, but that ultimately, it is the interpretation of the information which is the key.

These insights are particularly relevant in a multi-stakeholder environment, such as that in South Africa, where interpretations of data often differ substantially. It is crucial to a deeper appreciation of water information systems management that we understand the central role that knowledge plays in the interpretation of information. Opinions are formed, standpoints adopted and actions taken only after the interpretation of information. It is thus, not the information per se, but access to it in relation to other relevant information and interpretation through an individual's or a group's knowledge, that is important.

11.3 Primary Sources of Observed Water-Related Data

In South Africa, a wide range of water-related data is observed by thousands of independent observers, who transmit this data to a lesser number of organisations for collation and storage. The data collectors include South African Weather Services,

Departments of Agriculture, Local Government, and private sector institutions. DWAF is a direct observer and primary collector of only a modest portion of water-related data in South Africa.

The aforementioned data are generally stored in a plethora of formats on computer and in paper filing systems in the organisations that carry out the primary collection. In large government departments such as DWAF, the Department of Agriculture, and DEAT, this data is often housed in separate data bases, even within the same department.

Thus, while DWAF is the custodian of the country's water resources, it is dependent on a large number of organisations for the primary observed data on which its water information management systems depend. These data collecting organisations also have data and information management systems which they use in their daily work. Yet, many of these organisations have little or no daily working relationships with each other, let alone a close information-sharing relationship. When one considers the essential steps required to quality check and transform primary observed data into useful, trusted information, one becomes even more aware of DWAF's dependence on other organisations. While a number of institutions have copies of primary-observed data collected by others, there is very little genuine integration of data, let alone information, despite numerous attempts in the past. These attempts have generally ended in failure due to internal resistance and lack of co-operation.

11.4 Generation of Information from Primary Observed Data

One of the universal challenges in the field of water management is making wise decisions based on limited, scattered or absent data. This challenge is particularly present in South Africa. Assumptions need to be made to fill in the gaps. If these assumptions are explicitly declared, and the formal path from data through assumptions and calculations to information is explicitly laid out, we call this a model. This is a key area of endeavour in the development and interpretation of water information for individuals. In South Africa, where diverse groups are often in contention over water information, an understanding of the progression from data-to-information-to-knowledge, and finally to wisdom, and particularly collective wisdom, is critical.

In a co-management setting there will be many different views and interpretations of data and information. Since modelling is one of the primary means of creating water-related information from observed data, it is important to take a close look at the process of modelling from perspectives that are not often discussed.

It is commonly stated that models are useful for generating *information* about water systems so that *options* can be considered and *decisions* taken to manage the water systems and *resolve* conflict. This description of the issues was adequate for the water supply era. However, South Africa is now in an era of equitable allocation and services provision. Water allocation and provision are essentially social processes, with technical elements. These social processes require:

- ***Credible*** information, ***trusted*** information, ***shared understanding*** of information
- ***Sensible*** options, ***innovative*** options
- ***Shared understanding of consequences*** of decisions, ***acceptable*** decisions, ***wise*** decisions
- ***Equitable*** resolution, ***peaceful*** resolution, ***lasting*** resolution, ***timely*** resolution

In a society with the history of disparity, inequity and mistrust of South Africa, the key question is how credibility, trust, shared understanding, sensibility, acceptability, wisdom, equity and peace can be achieved. The processes which yield these types of information, options, decisions and resolutions, are not based on creating information through uni-disciplinary, uni-organisational models, and uni-laterally disseminating the information, with little prospect of receiving and responding to feedback. They will be processes which:

- Offer regular, affordable and meaningful communication amongst all stakeholders
- Are flexible and iterative
- Increasingly reveal more information on the system dynamics
- Are open and transparent
- Enable implicit assumptions and mental models to be made explicit
- Foster generative leadership, as described by Senge (1990), and adaptive management
- Incorporate and reflect the input of all stakeholders
- Involve a form of integrated systems information generation, which can function in a data-poor environment
- Overcome the barriers to communication between stakeholders which arise from geographic, capacity, disciplinary and organisational separation

This list illustrates why it is so important to have democratic water governance institutions. Water resources and water services allocation are both deeply social processes, which need a solid scientific underpinning and a mature acknowledgement of the subjectivity of the information being used.

Implicit in all these processes is the requirement to develop the skills and technology that will enable the phenomena of inference, connectivity, credibility, trust, assumptions, perceptions, relationships and co-ordination, to flourish in the management of the information. All these phenomena, and especially that of co-ordination between groups, are social processes which require extensive practice to manage effectively. Well-designed information systems, especially those containing information-generation software (simulation models), require practise and feedback in their development and wise application. All the systems must also link into the systems which the major government departments have developed. In 1998, a DWAF/UNESCO/WMO report made strong recommendations for the development of the skills described above, and the then Minister of Water Affairs and Forestry, in the foreword of that report, said:

> The sustainable development and management of South Africa's limited water resources, so eloquently expressed as "some for all, forever", will not be achievable without the drive and support of a professionally equipped and dedicated cadre of engineers, scientists, technologists, academics, managers, administrators and auxiliary personnel, at all levels... This is an issue of immense proportions that requires immediate actions and consistent momentum (Asmal 1998).

These words describe the ideal for which South Africa is striving. The organisational forms to achieve this have been legislated and, in some cases, formed. However, behaviours have not yet changed to match the required needs of integrated water management. There has, therefore, not been a strong and consistent demand for integrated information either within DWAF or more broadly within government. Often water-related analysis work is outsourced to consultants, often with a Terms of Reference lacking an integrated perspective. The project specifications may require the time-consuming task of gathering relevant data from a range of organisations, transforming the data into formats which suit the proprietary systems of the particular consulting firm, checking the data, infilling gaps (e.g. in point rainfall observations) and using the data and information for the required analysis. After completion of the project, the data collection and the systems to store and manipulate it too often remain with the consultant rather than DWAF. There is often a lack of a pre-designated "home" for storage of the data or information in DWAF, and the formats of the stored data and information, and the programs to generate the information, access, and display it, are often linked to proprietary formats of the different firms. The sets of data and information are of little value without the systems to access and manipulate them, make quality and source statements about them, and detail the assumptions and techniques that went into creating information from the data. The net result is that there is a plethora of data formats, generated information (e.g. infilled rainfall) and manipulation programs stored in data and information collections in numerous consulting firms, research and academic departments throughout South Africa.

11.5 Key Operational Requirements of Water Information Management Systems

The preceding sections have focused mainly on the strategic socio-scientific imperatives of the water information management challenge. It is reasonable to conclude that, whilst water stakeholders in South Africa have begun to grasp these imperatives, they are still some way off implementing the operational actions that such imperatives require. This section addresses the key operational aspects which water industry stakeholders need to consider. Interestingly, in the field of information technology, matters which may, at first, appear to fall into the operational category are, in fact, strategic. Consider the matter of inter-operability standards, such as internet data transmission protocols or USB communication port data protocol standards. A key operationally strategic question is what the inter-operability standards in the water information systems field are or should be.

11.5.1 Inter-Operability Standards

Water (rainfall, rivers, and groundwater) does not respect human-made organisational, disciplinary or political boundaries. These boundaries are artificial constructs which humans create and which, in many cases, hamper the wise management of our behaviour with regard to water. The resultant organisational fragmentation of water-related information presents barriers to the development of integrated water information systems. One of the fundamental skills required of practitioners, in this area of endeavour, is the ability to transcend boundaries, in a wise manner. In many areas of human endeavour, this challenge is present and it is addressed, inter alia, by the adoption of de facto inter-operability standards to effect exchanges at the boundaries. De facto standards are ones that are adopted out of choice and not because of a legal requirement – one adopts such standards because of the benefits offered.

The broad realm of water management has a great need to develop inter-operability standards for data and information exchange between a wide range of institutions, including DWAF, CMAs, WUAs, SALGA and so on. The stakeholders and managers in these organisations will, inevitably, have to interface their information flows with one another and work together if their respective plans are to be logically consistent, integrated and coherent. Catchment Management Strategies (CMSs), Integrated Development Plans (IDPs), Environmental Management Plans (EMPs) must all integrate and be consistent with the Spatial Development Frameworks (SDFs) and Environmental Management Frameworks (EMFs). Sharing of data, information, knowledge, common information systems and even personnel, is a fundamental pre-condition for coping in such a complex, dynamic and interdependent world.

Thought leaders in South Africa are encouraging water scientists and engineers to learn information integration lessons from other industries. They are pointing to the value of integrated interfacing and the sharing of common platforms and inter-operability standards which have been demonstrated by the computer, financial services, telecommunications, aerospace and even the retail industries (where bar-coding is an example). These thought leaders are pointing to de-facto inter-operability standards in the field of water which are emerging elsewhere in the world, where markets are huge. Inter-operability standards are driven largely by market numbers and so South Africa will be a follower rather than a leader in setting such standards. For South Africa, the benefits of adopting de facto inter-operability standards are substantial, not least for systemic innovation which requires a delicate balance between control and the intellectual and technological space within which to innovate. Chesbrough and Teece (1996), explain systemic innovation as one whose benefits can only be realised in conjunction with related complementary innovations. South African scientists and engineers are beginning to recognise that this is an important principle for designers of water information generation software (models) who intend contributing to integrated water management. Chesbrough and Teece (1996) believe that systemic innovations pose a unique set of management challenges regarding information exchange, sharing and co-ordinated adjustment throughout the entire product system, and often the organisation/s involved. They maintain that co-ordinating a system innovation is particularly difficult when industry inter-operability standards do not exist and must be pioneered. Increasing numbers of South African water engineers and scientists are recognising the validity of this assertion.

According to Chesbrough and Teece (1996), experience has shown that, once a new inter-operability standard has been established, virtual organisations can manage further innovation quite well. Inter-operability standards should not be confused with quality standards. These comments, relating to systemic innovation, are supported by Upton and MCaffee (1996)), who reported on a number of innovative, virtual factories which link across networks to have physical components manufactured at remote sites. The need to build and nurture networks of scientists, social scientists, economists and engineers in the service of water is implicit in the policies and legislation which emerged in South Africa post-1994. More recently, this has been re-emphasised with a programme called the Framework for Education, Training and Research in Water (FETWater). The programme stems from an extensive investigation into the education and training needs in the field of water, launched by the then Minister of Water Affairs & Forestry, Professor Kader Asmal in 1998 (DWAF/UNESCO/WMO 1998).

Perhaps the best-known examples of inter-operability standards are in the computer industry where plug and play compatibility is now taken for granted and where the Internet is the pinnacle of inter-operability standards achievement. Ironically, we all use the Internet daily and yet, the South African water sector is not transferring its most powerful lessons – the value of inter-operability standards – to the field of water information management. One example of this is the lack of adoption of the public domain, open source, ArcGIS Hydro data structures or the BASINS 4 framework, which include time series data standards. These will be discussed briefly later in the chapter.

11.5.2 Access to Data and Information

Convenient, unimpeded and equitable access to data and information is fundamental to the democratic process of water allocation and delivery. At present, in South Africa, access to information is none of these. For example, there is not one computer system in South Africa onto which water data users and information creators from different organisations can routinely log on and genuinely work together. Firewalls, red tape, unhealthy competition, and limited capacity prevent such minimum levels of co-operation. So, whilst it is possible to access copies of data sets, the organisational fragmentation is severe and the barriers to co-operation and integration are high and strong.

CMAs will be the first, and only, institutional forms in the water resource management sector specifically designed to encompass all stakeholders in a permanent, transparent, accessible and integrated manner. Registered users from all stakeholder sectors should have a legitimate right to logon to the CMA computer systems and contribute to the development and use of their data and information systems. All water users will be paying for these systems and, therefore, it is inconceivable that they will be prevented from using them. This will herald the start of a whole new way of working which will take co-operation and genuine information and modelling systems integration a giant step forward. The manner in which stakeholders interact in and through CMAs will, undoubtedly, transform the development and management of water information and hence water management, fundamentally. However, key questions, relating to the digital and knowledge divide between rich and poor, need to be addressed.

11.5.3 *Affordability of Information*

Affordability of information is a key and complex issue. Affordability is linked to value, and the utility value of information is determined by the user of the information and the timing of the need. Utility value must not be confused with cost, which is the total cost of observing the data or deriving the information, divided by the number of times that item is used. The unit cost of data or information, therefore, decreases every time it is used. The use value reflects the level of trust and confidence that one has in the item before using it. This, in turn, also depends on the alternatives that one has. As the saying goes, "in the land of the blind the one-eyed man is king". The value and benefit of a single good rainfall station on the top of a mountain in Lesotho is considerably more than one in a uniform rainfall area where it is accompanied by ten other rain gauges in a 1 km^2 area. The value, benefits, costs and selling price all combine to form the notion affordability.

In South Africa, affordability is a key element in the equity debate and in the social process of the allocation of water services and resources since affordability also influences access. Data and information may be physically accessible but not affordable in monetary terms. Similarly, data and information may not be affordable in terms of the cost of converting it into useful knowledge. This introduces the all-important factor of expertise or knowledge power that is discussed briefly later.

The concepts of economies of scale and sunk costs are also important to understand in the context of water information. The information communication and technology (ICT) realm is one in which economies of scale have been well-exploited. Inter-operability standards have played a key role in developing such economies of scale. The concept of sunk costs and, particularly the protection of sunk costs and their unit reduction by using economies of scale, is key to the success of the computer industry.

11.6 Changing Roles of Water Information Developers and Managers

As the engagement between water stakeholders transforms, the role of water information developers and managers also changes. The increasing complexity of water problems and the democratisation of the processes to resolve these problems are key drivers of this changing role. A central challenge in the multi-stakeholder water management environment in South Africa is to generate information that is credible, trusted, understandable, timely, and shared, in an environment that is historically skewed in terms of knowledge power. To understand the unequal knowledge power, we need to consider who the water information developers and managers are and which stakeholders they advise.

The White Paper on a National Water Policy for South Africa (RSA 1997) and the National Water Act (RSA 1998) stress the importance of equity in access to water and in water management institutions such as CMAs. Equitable representation, however, is not enough in the complex, dynamic field and multi-disciplinary realm of water, where knowledge plays such a strong role. Access to information is also not enough to ensure equity. The real power lies in knowledge, and in this respect, the playing field is skewed strongly in favour of the white minority.

11.6.1 Unequal Knowledge Power

Twenty-five years ago, most of the water knowledge power in South Africa resided in state departments. Such a knowledge power setting was adequate to cope with the "get more water" and "use water more efficiently" eras. However, South Africa is now in the "water allocation era" and significant knowledge power resides with powerful water users that have employed increasing numbers of water experts in recent years. This raises strategic questions about the lack of knowledge power amongst the marginalised and the vulnerable and what, if anything, South Africa has done or can do to try to address this. This shift in the balance of knowledge power holds important strategic implications for the development and use of integrated water resource information generation systems (i.e. models) which are used in the social process of water allocation. It is increasingly recognised that the state has a duty to ensure knowledge power equity amongst stakeholders. The new bargaining paradigms which are likely to emerge and which are outlined later in the chapter, will assist the state in this task. In addition, opposing contentions amongst the stakeholders over the quality and quantity of the water resources may help to ensure the checks and balances required to restrain any potential abuse of knowledge power. Leading thinkers in South Africa are increasingly reflecting on the value of the institutional and legislative platforms that the NWA presents for dealing with these imbalances wisely.

11.6.2 Managing Data and Information in a Multiple Ownership Environment

In a democratic multi-stakeholder setting, affordable access to data and information for all is non-negotiable. This begs the question then, of who pays for information. This question is linked to cost, value, benefits and affordability. It also links to the questions of who gets paid, and who owns the data and information. There are ongoing debates over who pays for water-related data and information, debates that have generally not been resolved and will not be sensibly resolved until there is a maturing of the paradigms within which they are framed. It is necessary to recall that the sources of primary observed data in South Africa are many and organisationally diverse. This is both a strength and a weakness. It is a strength because it produces a healthy diversity in terms of data ownership power. It is a weakness because it introduces an additional level of organisational and technological complexity into the production of useful information.

It is particularly interesting to consider the joint ownership of the information which will emerge from joint stakeholder modelling processes in multi-stakeholder institutions. One of the key attributes of such information, that will set it apart from other information, is that it will carry a mantle of trust, credibility and buy-in bestowed by the process. Currently, much of the information on which decisions are being made is generated by consultants under contracts in which they and DWAF or water services authorities, are the only participants. This unhealthy state of affairs will end when consultants, working on projects for CMAs, become aware that they are employed by the stakeholders that are contributing to CMA costs. However, implementation is slow and it is evident that several key role players are reluctant to see their former powers curtailed as they engage the democratic process which will bestow a mantle of approval on the generated information.

Information that is not part of the joint process will not carry a significant mantle of approval and will, therefore, have far less value. This assertion is supported by Reitsma et al. (1996), when they consider the subjective nature of models which generate information for use in water negotiations. They report that the sharing of models and information among interest groups assumes the acceptance by all parties of those models. They state that, at first, this may seem straightforward and non-problematic since models are intended to represent the objective properties of the natural resource. However, since water information generation systems are the product of human thought and are, in essence, a sequence of assumptions, they typically have a subjective cultural background. In addition, they are often developed within groups or organisations that also take part in the negotiation process, either as parties or as external domain experts. They conclude with a strong statement, that a careful study of the role of simulation models in water resource negotiations requires analysis of a number of strategic, tactical and managerial aspects of using generated information that emerges from water simulation model use. The issues they raise are particularly relevant in South Africa where actual observations are sparse and assumptions within models play a large role in the generated information which often forms the centre of contentious water debates.

The development and use of such models provides a framework for thinking and double-loop learning, which Liedtka et al. (1997), explain as requiring a fundamental re-examination of the assumptions underlying information generation. The model assumptions influence the generated results, and hence, the reasoning that stakeholders engage in, based on the model output. Since water quality and quantity systems are complex, the modelling systems must be able to reflect this complexity.

Highly skilled and knowledgeable people are, therefore, required to understand and use these models in the collective bargaining setting that will characterise the CMA. It is, therefore, important that all stakeholder groupings have knowledgeable and trusted professionals to advise them in these complex matters.

The following section focuses on, arguably, two of the most crucial aspects of water information systems management, namely, water bargaining paradigms and knowledge equity. Both these require a focus on the organic emergence of innovative ways that expertise is organising itself to serve water resource management institutions. It is an area in which government is increasingly conscious of the dynamics, and senior officials are beginning to speak of interventions to enhance interest-based bargaining and knowledge equity.

11.6.3 Water Bargaining Paradigms and Knowledge Equity

In the water services sector, the issues around information and knowledge equity are large and complex. The pathway to resolving these is perhaps not as clear as that in the water resources realm. However, the new paradigms which are emerging more strongly in the latter will, inevitably, create a more fertile environment for solutions to emerge in the water services sector. Two of the key areas are water bargaining paradigms and knowledge equity.

A key to understanding this paradigm shift is the recognition that the allocation of entitlements to use and abuse (pollute) water is, essentially, a social process of bargaining. Broadly speaking, such processes can take on the attributes of either rights-based or interest-based bargaining. It is important to understand these two processes, which are well-documented in the industrial stakeholder relations literature and, increasingly, in the natural resources management literature, for example Hrezo and Hrezo (1985); Bruns (2003); Spangler (2003); Bruns (2005); Van Wyk et al. (2006). The behaviour of the parties in each paradigm is markedly different, with implications for the human, financial, legal and technological resources which are utilised by the process. All of these resources are controlled by the behaviour of the parties and the nature of the bargaining paradigm has a major effect on many aspects of the process. This is one of the issues that Quadir et al. (1999) were referring to in this statement:

> The impacts of information technology on the water sector are not inherent in the technology but largely depend on the way society chooses to use the technology. The new technology does offer unprecedented possibilities to change knowledge relationships which impact on power relationships and consequently on organisations and society at large (Quadir et al. 1999).

Astute leaders will strive to ensure that the interest-based process is dominant although the first impulsive choice of many may be the rights-based bargaining paradigm and the introduction of lawyers and the courts into the process. A glance at the contrasted characteristics of these two processes will reveal the wisdom of developing leaders rather than paying lawyers (Table 11.1).

Table 11.1 Rights based versus Interest based bargaining

Rights based bargaining (lawyers)	Interest based bargaining (leaders)
Appropriate when only two parties and uncertainty is low	Appropriate when more parties and uncertainty is high
Adversarial process	Co-operative process
Stronger seeks to dominate the weaker	Stronger seeks to empower the weaker
No sharing of information	Open sharing of information
Large scale duplication of effort	Productivity enhanced through sharing effort
Difficult to reach decisions in uncertain environments	Easier to reach decisions in uncertain environments
Expensive in terms of intellectual resources	Effective use of intellectual resources
State has difficulty ensuring a fair deal for the disadvantaged	State can leverage voluntary help of stakeholders to empower disadvantaged
Poaching of skills encouraged	Reduced incentive to poach skills

The White Paper on a National Water Policy for South Africa (RSA 1997), the Water Law Principles, the National Water Act (RSA 1998) and the DWAF Strategic Plan March 2000 (DWAF 2000), all imply that the interest-based paradigm will be encouraged in water allocation processes. Unfortunately, to date, the potential inherent in interest-based bargaining has not been realised for several reasons. Firstly, the insights of the drafters of the legislation have not been matched with the ability of DWAF to let go in the implementation. Significant sections of DWAF have been openly sceptical about CMAs, which partly explains the 10 years that have passed before the first CMAs are beginning to function. Secondly, significant role players among white water users and the consulting sector have been playing a very astute strategic game in slowing the implementation of the NWA, whilst not appearing to be openly opposed to its implementation – the NWA is fundamentally about redress in access to water resources and it would be naive to believe that all stakeholders welcome it. Thirdly, there have been well-meaning role players, within and outside DWAF, who believe that the interests of the poor are best served by strong central government control. The flaw in their reasoning has, however, been that much expertise has left DWAF to join the consulting sector and well-resourced stakeholders in the private sector, leaving government officials being "run off their feet" by the complexity and extent of issues. As a result, the status quo remains with respect to water allocation.

Against this background, there are a growing number of voices pointing to the benefits of interest-based bargaining, which they believe will steer participants towards a power dynamic which uses information as a light rather than as a weapon to win unfair advantage. It is argued that interest-based bargaining will help to close the knowledge power gaps, but the process must be assisted by the state financing trusted expert advisors for poor and disadvantaged stakeholder sectors. The resources of state departments other than DWAF can be utilised to provide advisors to the previously disadvantaged. Institutions such as the Council for Scientific and Industrial Research (CSIR) and the Agricultural Research Council (ARC) could fulfil the role of pro deo advisors to the poor and weak. The state could also pay consulting groups to fulfil this role. Perhaps well-resourced stakeholders will, themselves, begin to see the wisdom of having well-advised negotiating partners. It has been the experience of such forms of negotiation in business that demands are much more sensible, and consensus is more likely, when all stakeholders are knowledgeable.

11.6.4 Institutional Memory Building for New Water Institutions

The building and retention of institutional memory is critical for the success of any complex multi-stakeholder endeavour. The socio-political and scientific complexity in multi-stakeholder CMAs makes the building of institutional memory a strategic imperative. The manner in which integrated information and modelling systems are developed and managed is central to these efforts. One of the unfortunate consequences of the inordinate delays in the implementation of multi-stakeholder water democracy in South Africa has been the continued fragmentation of information and the dissolution of institutional memory, combined with the continued migration of expertise away from DWAF – a trend that was noted almost 10 years ago in the UNESCO/WMO/DWAF (1998) report. There has also been a movement of key information systems, particularly those that involve simulation modelling, away from DWAF. Fortunately, this trend of increasing fragmentation may have reached its zenith as CMAs begin to form a focal point for the re-integration of information and information generation systems. As the role players become more accustomed to the imperative for sharing and the transparency of data and information systems, so the search will intensify for integrated systems that will serve as a focal point for stakeholder interaction. These integrated systems will contribute to better institutional memory building. As can be anticipated, there are many contenders putting forward the systems that they have developed in a bid to place them in a powerful position as a preferred supplier of the core integrated information and modelling systems platform for CMAs. Clearly, capturing this market puts one in a powerful business position. This, however, would be unhealthy for the water democracy dynamic, and would inhibit the development of a sound, broad-based institutional memory.

It is reasonable to assume that open source, public domain systems that have been developed, tested and extensively used within a multi-stakeholder environment could be considered for use in South Africa. Increasingly, role players in South Africa are recognising that the widespread adoption of such systems will mean that no matter what catchment, area or organisation a particular expert is representing, they will be contributing to a common knowledge and institutional memory base. This has huge advantages for wise water management which is heavily reliant on good institutional memory. Given the enormity of the challenges and the paucity of water-related skills in South Africa, it is evident that this strategic pathway makes sense.

11.7 The Role of Information in Co-Operation and Feedback

While information is key in the co-operation and feedback which is so vital to service delivery, it often does not effectively lubricate co-operation and service delivery. If information is created in a fragmented and secretive manner, which invites distrust, it can inhibit co-operation and even cause conflict. It is not information per se that is useful, but information in combination with a healthy process of information generation. The potential to engage such a healthy process in South Africa has been improved greatly since 1994.

Co-operation is generally enhanced by trust between stakeholders. South Africa has a long history of domination and mistrust and it would be naïve to assume that this will be replaced by trust and co-operation overnight. The development of trust is a slow process which can be enhanced by stakeholder involvement in the generation of information. Interest-based bargaining and the knowledge power dynamic have been discussed briefly above. A further key element in the process is the feedback from modelling systems and the modellers' responses to the feedback. The Fig. 11.3 below provides a rough schematic of the steps in the data gathering and modelling process which includes feedback from the models that can guide monitoring and indicate the sensitivity of the overall system to data input changes. Such

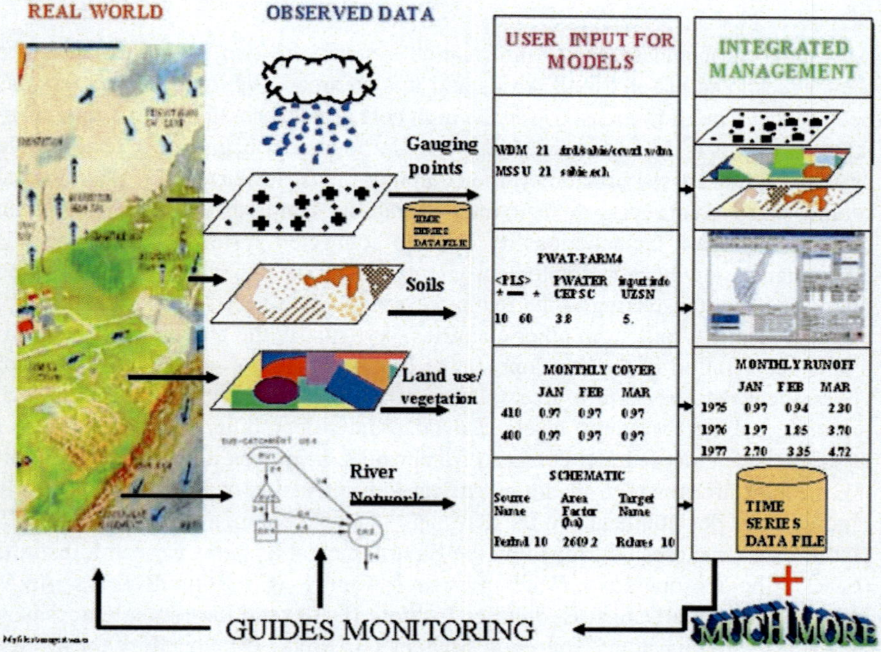

Fig. 11.3 Conceptual pathway of data from field observations to model input and feedback to guide data collection

sensitivity analyses are useful in determining the location, frequency and accuracy required of monitoring systems in the catchment.

The discussion, in the sections above, leads to the recognition of a role for information that is not based on the slogan, "Information is Power". The increasing realisation of the complexity of integrated water resources management issues has fed the development of what are often referred to as Decision Support Tools.

Wise leaders in South Africa are increasingly sharing in conversations which recognise the requirement for participants to ***reason*** together, to think together, to explore consequences together and to reflect systemically together. Naturally, decisions will have to be taken at the appropriate time. However, the information management and generation systems that we commonly refer to as ***decision*** support systems are primarily useful in the *integrated reasoning* stage. Information generation systems enable us to declare our assumptions and, in the terminology of dialogue, to suspend our thoughts in an open space for all to see and reflect on. Such systems are a core aid to developing institutional memory and deeper levels of learning explained in terms of the U-diagram by Senge et al. (2005). It is not surprising to witness the emergence of the term, "integrated reasoning support systems" rather than decision support tools to describe information and information generation systems.

11.8 Converging Trends Driving Affordability and Access

Affordability of and access to information systems are two potential barriers to their widespread use in South Africa and it is an area in which little progress has been made. One of the reasons for the high cost and inaccessibility of some information systems is the high transaction costs introduced by using non-inter-operable systems. As discussed previously, the computer industries have made phenomenal progress through applying de facto inter-operability standards to reduce transaction costs and increase the functionality of large connected systems. Underlying this phenomenal growth are technological and organisational behavioural lessons which many are now transferring to the realms of water, albeit very slowly in the case of South Africa. Lessons from other countries, particularly the USA, point to considerable acceleration in the development of low transaction cost routes from major government data bases to usable systems at the multi-stakeholder modelling and reasoning support systems levels. De facto inter-operability standards such as OpenMI have emerged and common frameworks have been developed to support this need. Unfortunately, in South Africa, key interest groups have consistently ignored these developments, to the detriment of all except themselves. The ArcGIS Hydro data structures developed by the Environmental Systems Research Institute (ESRI), the Co-operative Water Research Centre (CWRC) at Texas A&M University and the Danish Hydraulics Institute (DHI), and the Better Assessment Science Integrating point Non-point Sources (BASINS) developed by the United States Environmental Protection Agency (US EPA), the United States Geological Survey (USGS), and AquaTerra, are two such products. These products have been used extensively, inter alia, in the Total Maximum Daily Load (TMDL) programme of the US EPA, which involves approximately 40,000 impaired catchments in the USA. These developments have shown that once the data structures at the catchment level of multi-stakeholder organisations have been fixed, then all the major data repositories in government departments can provide extraction and reformatting routines, and make these available over the Internet. Set-up times to extract numerous data sets and populate reasoning support systems for multi-stakeholder catchment organisations have been transformed from months to hours. Access and affordability have improved dramatically.

These systems are in the public domain and are gaining increasing attention in South Africa, which has been inappropriately slow in learning and adapting to new technology and approaches in information management and systems.

11.9 Conclusion

This chapter has shared some of the challenges which South Africa is facing with regard to water-related data and information in the context of a digital divide, severe human resource and financing constraints, and severely skewed knowledge power in a multi-stakeholder environment. Also addressed in this chapter, are some of the lessons from South African and international experiences.

A key lesson is that the ultimate purpose of data collection is to result in action. Data that is not integral to a process that gets people to act is useless. However, data per se does not have the power to get people to act. It is only as data becomes transformed by the human social process of understanding and relating in a connected system that it becomes the fuel that moves people to action. Just as food needs to be ingested and digested before it transforms into the appropriate chemical state to energise our bodies, so it is with data. The societal body must ingest and digest water data and information before it can be understood and trusted enough to move the societal body to wise action. The often robust process of formulating collective understanding of information generates what Nowotny et al. (2001) refer to as socially robust knowledge. Knowledge needs to be socially robust for it to engender action in society (Nowotny et al. 2001). The complexity of this societal body in South Africa should be evident to readers of this book. All parts of this body must have access to the data and information and must have the collective capacity to understand it and trust in the processes which generate it, before wise actions can be agreed upon. This is not easy as Wheatley (2006:99) remarks, concerning organisational dynamics in general,

> ….it is taking us some time to learn how to live in open intelligent organisations. This requires an entirely new relationship with information, one in which we embrace its living properties. Not so that we open ourselves to indiscriminate chaos, but so that we can facilitate effective responses in a world that is constantly surprising us. If we are seeking resilient organisations, a prized property of living systems, information is a key ally.

The creation of a monitoring and information management system is a necessary, but not sufficient, condition for the meaningful societal participation process which precedes wise action. The sociology of information creation, ingestion and digestion, in a multi-stakeholder environment, must be appreciated before any group can specify a satisfactory system. This is perhaps the key lesson from the South African experience and, paradoxically, it is a key lesson that South Africa still needs to learn from international experiences in the water and other sectors. The full potential of South Africa's excellent policy, legislation and institutional platform needs to be realised through wise approaches to data, information and knowledge. Achieving this will make South Africa a centre of learning for the rest of the world.

References

Asmal K (1998) Water as a metaphor for governance: issues in the management of water resources in Africa. Water Policy 1:95–101

Bruns B (2003) Water tenure reform: developing an extended ladder of participation "Politics of the commons: articulating development and strengthening local practices." Paper presented at the RCSD Conference, Chiang Mai, Thailand

Bruns B (2005) Community-based principles for negotiating water rights: some conjectures on assumptions and priorities. Paper presented at the international workshop on African water laws: plural legislative frameworks for rural water management in Africa, Johannesburg, South Africa

Chesbrough HW, Teece DJ (1996) Organising for innovation. Integrated virtual alliances. When is virtual virtuous? Harv Bus Rev Jan–Feb:65–73

DWAF (Department of Water Affairs & Forestry) (2000) Department of water affairs & forestry strategic plan 2000. Department of Water Affairs and Forestry, Pretoria

DWAF, UNESCO, WMO (Department of Water Affairs & Forestry, United National Education, Scientific & Cultural Organisation, World Meteorogical Organisation) (1998) Mission on the assessment of the education and training needs of the water resources management services of the Republic of South Africa. DWAF, UNESCO, WMO, Pretoria. ISBN 0-620-22756-7

Hrezo MS, Hrezo WE (1985, April) From antagonistic to co-operative federalism in water resources development: a model for reconciling Federal, State and Local programs, Policies and Planning. Am J Econ Sociol 44(2)

Liedtka JM, Haskins ME, Rosenblum JW, Weber J (1997) The generative cycle: linking knowledge and relationships. Sloan Manage Rev (Fall)

Nowotny H, Scott P, Gibbons M (2001) Re-thinking science knowledge and the public in an age of uncertainty. Polity Press, Cambridge, UK

Quadir G, Dueck M, Abbott F, El-Baz K, Takeuchi H, Phien N (1999) Report of the thematic panel on information and communication technology and its implications for water resources. World Water Vision. International Water & Sanitation Centre, The Hague and the Swedish International Development Agency

Reitsma R, Zigurs I, Lewis C, Wilson V, Sloane A (1996) Experimentation with simulation models in water-resources negotiations. Water Res Plan Manage Jan–Feb:64–70

RSA (1997) White paper on a national water policy for South Africa. Government Printer, Pretoria

RSA (1998) National Water Act. Government Gazette Vol. 398. 26 August 1998. Office of the President, Cape Town

Senge P (1990) The leader's new work: building learning organisations. Sloan Manage Rev 32(1)

Senge P, Scharmer CO, Jaworski J, Flowers BS (2005) Presence, exploring profound change in people, organizations and society. Nicholas Brealey Publishing, London

Senge P, Roberts C, Ross RB, Smith BJ, Kleiner A (1995) The fifth discipline field book: – strategies and tools for building a learning organisation. Nicholas Brealey, London

Spangler B (2003) Integrative or interest based bargaining. Knowledge base essay in Beyond intractability. http://www.beyondintractability.org/essay/interest-based_bargaining/. Accessed 18 May 2007

Sterman J (1989) Misperception of feedback in dynamic decision making. Organ Behav Hum Decis Process 43(3):301

Turton AR (1999) Water scarcity and social adaptive capacity: towards an understanding of the social dynamics of managing water scarcity in developing countries. MEWREW Occasional Paper 9, Water Issues Study Group, Department of Geography, SOAS University of London

Upton DM, MCaffee (1996) The real virtual factory. Harv Busi Rev July–August:123–133

Van Wyk E, Breen CM, Roux DJ, Rogers KH, Sherwill T, van Wilgen BW (2006, July) The ecological reserve: towards a common understanding for river management in South Africa. Water SA 32(3). ISSN 0378-4738

Wheatley MJ (2006) Leadership and the new science: – discovering order in a chaotic world. Berrett-Koehler Publishers Inc., San Francisco, CA

Chapter 12
The Water Research Commission

Jayant Bhagwan

Abstract The South African water sector endeavours to ensure that scarce water resources are, and will continue to be, managed and used optimally and sustainably for the benefit of all citizens. Given the numerous and often serious challenges faced by the sector, strong support is needed in the area of research and development (R&D).

Since 1971, water-related research and development on behalf of the government and water sector of South Africa has been vested in the Water Research Commission (WRC). Funded in terms of the Water Research Act of 1971 through a levy on water use, the WRC reports to Parliament through the Minister of Water Affairs and Forestry. The WRC commissions research to address identified needs and disseminates resulting knowledge with a view to solving water-related problems. The WRC also plays a key role in capacity building in the water sector.

Keywords Water management • Water policy • Research and development • Impact • Knowledge generation

12.1 Introduction

This chapter discusses the role of the Water Research Commission in funding research on water and how the policy and strategy of funding research on water has evolved over time, and was influenced by post-1994 policy and legislation.

Prior to 1971, water research in South Africa was largely unco-ordinated. Dedicated water research focused mainly on waste-water reclamation and was limited to a single research organisation, the National Water Research Institute (NWRI). In accepting the recommendations contained in the 1970 report of the Commission

J. Bhagwan (✉)
Water Research Commission, Private Bag X03, Gezina 0031, Pretoria, South Africa
e-mail: jayb@wrc.org.za

B. Schreiner and R. Hassan (eds.), *Transforming Water Management in South Africa: Designing and Implementing a New Policy Framework*, Global Issues in Water Policy 2, DOI 10.1007/978-90-481-9367-7_12, © Springer Science+Business Media B.V. 2011

of Enquiry into Water Matters (RSA 1970), the government acknowledged the important role of research in water management in South Africa, and adopted a policy of intensifying and co-ordinating water research at the national level.

This policy gave rise to the Water Research Act of 1971 (RSA 1971), which established the WRC and gave it the mandate to co-ordinate and support water research, and to disseminate and encourage the application of research results, using funds from a dedicated water research fund (WRF) created for this purpose. The WRF's main sources of income are levies based on metered quantities of raw water sold, with the exception of irrigators, who pay a levy on the basis of the area (hectares) of irrigated land. This created a unique situation of water research funded through a research levy as part of raw water tariffs.

Thus, since 1971, the responsibility for leading water-related research and development (R&D) on behalf of the government and water sector has vested in the WRC. The WRC is governed by a Board of Commissioners, appointed by the Minister of Water Affairs and Forestry, and reports to the Minister through the Board. The WRC does not do research, but assesses and prioritises research needs in close consultation with stakeholders, contracts research to meet these needs and disseminates resulting knowledge with a view to solving water-related problems. Inherent in the WRC's funding policy is the development of the capacity and skills needed to empower the water sector to support national growth and development objectives.

This situation does not imply that water research was solely vested in the WRC, as other institutions, government departments and science councils have been investing in water research. The strategic objective in the creation of the WRC was to ensure a dedicated platform for research and good science which ensured a continuum in the development of water knowledge, infrastructure, solutions and capacity. The WRC and the Water Research Fund (WRF) have ensured excellence in many water knowledge areas over the years and have brought significant benefits to the country and its water challenges.

The Water Services Act (RSA 1997a) and the National Water Act (RSA 1998), in giving effect to newly-developed national water policy, created new institutions and spelled out new water management objectives. The task of implementing the legislation posed new challenges to research and greatly influenced the national research agenda. The National Water Resource Strategy (DWAF 2005), among other things, addresses the national need for, and provides direction to, water-related R&D. With the issuing of an umbrella R&D Strategy by the Department of Science and Technology in August 2002, water research was, for the first time, placed within a broader national R&D context, and new opportunities for synergy were created. The Department of Science and Technology fulfils a national facilitating and co-ordinating role for all R&D and, thereby, serves to link water-centred R&D to all other relevant R&D initiatives in South Africa.

Investment in R&D targets the entire knowledge cycle, namely, knowledge creation, assimilation, dissemination, application and finally, innovation. It is only through the application of knowledge and innovation that return on investment in R&D is ultimately realised. The sharing and dissemination of new knowledge, created through water-centred R&D and the active encouragement of its application, is an ongoing initiative which the WRC maintains on behalf of the South African water sector as a whole.

The Department of Water Affairs and Forestry (DWAF) regularly commissions and funds investigations of direct concern to its day-to-day operations, but does not, itself, initiate research of a more strategic nature. However, since most water sector research will ultimately be of interest or use to DWAF, DWAF's priorities contribute substantially to the setting of the national water research agenda.

Water research is also initiated and funded by other organisations that have a strategic interest in doing so. The Council for Scientific and Industrial Research (CSIR), a member of the family of Science Councils and a provider of contract research services, may allocate its own funds to water research in the interests of building internal capacity and positioning itself to be more competitive in the market place. The same applies to other science councils, such as the Agricultural Research Council, the Council for Geosciences, the Human Sciences Research Council and the Medical Research Council.

The National Research Foundation (NRF) has a major responsibility for providing research funding support to tertiary educational institutions (Universities and Institutes of Technology or Technikons), in the interests of student training and capacity building. The NRF may also provide funding over a broader front (through vehicles such as the Innovation Fund) for research projects relating to national priority areas. Some of these projects may include facets of water research.

As is the case with DWAF, certain of the larger water boards and waste utilities (such as Rand Water, Umgeni Water and ERWAT), and industries (e.g. mining, power generation, petrochemical, textile), as well as commodity-based agriculture, may fund research related to their own areas of concern. Water-related issues are often addressed as part of such industry-focused research initiatives.

Since neither the WRC nor DWAF are performers of research, the water sector relies heavily on external providers (with exceptions in the case of Rand Water, Umgeni Water and ERWAT) that possess the necessary competence, skills and infrastructure to perform the research. The main research providers to the water sector are tertiary institutions (predominantly universities), and to a lesser extent, the science councils, research consultants in private practice, water utilities and NGO's. The relative contribution of the different categories of research provider to the national water research effort will be addressed later in this report.

Water-research providers in South Africa are, to a considerable degree, reliant on WRC support. Funding from non-WRC sources (at times substantial) often tends to be supplementary and aimed at extending the scope of WRC-supported research in a direction required by a specific client. Base funding from the WRC has been instrumental in enhancing the capacity of R&D providers to serve the water sector and the country as a whole. R&D undertaken is intended mainly to produce results for the public domain, but there is also a focus on R&D which has the potential of producing commercialisable intellectual property and R&D in which specific industries have an interest and which, therefore, enjoy their partial support (e.g. the mining and energy sectors). At the other end of the spectrum, information on industrial water-related R&D of a proprietary nature, largely self-financed by industry and often done in house (e.g. by the petrochemical and paper/pulp industries), is seldom reported, and does not directly benefit the broader water sector. As a result, this information is excluded from this chapter.

12.2 Changing Strategies for Water Research Post-1994

Prior to 1990, research funding was biased towards seeking high-tech solutions to "first-world" problems of water supply, ecosystem protection, treatment of potable and waste water and water-use efficiency in the municipal, industrial, mining and agricultural sectors. Conscious efforts to reduce this bias emerged from 1990 and accelerated rapidly after 1994, when ministerial approval was obtained for substantial budget enhancements to promote water research at disadvantaged institutions (technikons and previously disadvantaged universities). This drive was accompanied by the recognition of the broader stakeholder base that needed to be consulted in the setting of research priorities.

Since 1997, besides maintaining focus on the need for scientific and technological innovation, a large part of the R&D portfolio has been dedicated to supporting the successful implementation of policy and legislation and other national commitments, such as the Millennium Development Goals (MDGs). In addition, the emphasis on R&D as a vehicle for capacity building, has been stronger than ever before.

Investment in R&D needs to be adequate to mobilise a sufficiently strong research base (inputs) which will ensure that new knowledge, capacity and skills (outcomes) will empower the water sector to meet national requirements for the supporting of sustainable growth and development (impacts). Typical impact areas relate to:

- Empowerment of the sector to refine and better implement policy and legislation in close alignment with broader national goals and strategies
- Enhanced ability to protect/conserve water ecosystems in the interests of sustainability
- Greater adequacy of water, of suitable quality, supplied and equitably distributed for required social and economic development
- Overcoming constraints to delivering basic water and associated health/hygiene services to people
- Adoption of methods/technologies developed for maximising the use/benefits (and the reduction of pollution potential) of water made available for domestic, municipal, industrial, agricultural and other purposes

12.3 Strategic Areas of Water-related Research and Development at the WRC

The research portfolio of the WRC is structured according to four key strategic areas: Water Resource Management; Water-linked Ecosystems; Water Use and Waste Management; and Water Utilisation in Agriculture. In addition, four cross-cutting domains which address issues of national importance, namely, Water and Society; Water and the Economy; Water and the Environment; and Water and Health, provide integrating frameworks across the key strategic areas for research programmes and projects within each area.

12.3.1 Water Resource Management

The strategic focus for research in this area is largely guided by the principles and objectives of the NWA (RSA 1998). The primary underlying principle is that water resources should be managed to achieve optimum long-term social and economic benefits for all, while maintaining a balance between protection of the environment and efficient utilisation. The implementation of water policy is supported by developing tools and technologies for comprehensive water resource assessments, and guidelines and decision-support systems to support decision-makers in achieving the equitable and efficient allocation of water resources among competing needs. The research puts emphasis on multidisciplinary approaches that consider social, environmental and economic factors together with physical factors within the context of integrated water resource management. The research supports both policy-making and provides guidance for policy implementation and the development of policy instruments.

12.3.2 Water-Linked Ecosystems

Water-linked ecosystems are defined as in-stream (fully aquatic), riparian (dependent on water stored in the river banks and linked to the river) and water-table dependent (dependent on a water table, but not on surface water).

Research in this area addresses the conservation of aquatic ecosystems to provide knowledge for their sustainable functioning and the ongoing provision of goods and services which ecosystems deliver. Research further focuses on the protection and sustainable utilisation of the aquatic environment and biota (in-stream, riparian and groundwater). This includes the research needs around the international conventions on environmental management (e.g. biodiversity), as well as human needs from the aquatic environment (e.g. sustainable management for equitable ecosystem resource utilisation, recreation and eco-tourism).

12.3.3 Water Use and Waste Management

Research in this area focuses mainly on the domestic, industrial and mining water sectors. It aims to proactively and effectively lead and support the advancement of technology, science, management and policies relevant to water supply, waste and effluent management. Managing water use and the waste released into the water environment is of paramount importance to ensure the sustainability of the resource and the activities relying on it and to protect human health.

The focus is on water supply and treatment technology serving the domestic (urban, rural, large and small) systems, as well as the industrial/commercial and mining sectors of our economy. This area also focuses on waste and effluent as well as reuse technologies that can support the municipal, mining and industrial sectors and improve management in these sectors with the aim of improving productivity and supporting economic growth, while minimising negative effects on human and environmental health.

12.3.4 Water Utilisation in Agriculture

The strategic focus in this area of R&D is on increasing the efficient use of water for the production of food, fibre, fuelwood and timber; ensuring sustainable water resource use; reducing poverty: and increasing the wealth of people dependent on water-based agriculture. The needs and requirements of present and future generations of subsistence, emergent and commercial farmers, are addressed through the creation and application of water-efficient production technologies, models and information systems within the following interrelated sub-sectors of agriculture, namely, irrigated agriculture, dryland agriculture, grasslands, woodlands and forestry, livestock watering, and aquaculture.

12.4 Expenditures on R&D by Research Area at National Level

In this section, we present estimates of investments in water-centred R&D in South Africa by strategic research areas. Tables 5.1–5.3 summarise characteristics of South African water-centred R&D projects based on a sample of 274 multi-year WRC research contracts, centred on 2004, but contained within the period 1999–2009.

Table 12.1 shows that, while Water Use & Waste Management (WU&WM) had the highest number of projects, it is the least cost area per project, whereas Water Utilisation in Agriculture (WUA) projects are the highest cost, and extend over longer time periods (almost 4 years per project). This reflects the nature of the research undertaken to address problems of relevance to the strategic research area. It is also clear from Table 12.2 that more than two thirds of all R&D investments (70%) go to WU&WM and Water Resource Management (WRM).

Universities and Technikons are clearly the major providers of water-related research, using more than 56% of the funding on 55% of the projects, followed

Table 12.1 Research projects providing support to the South African water sector, based on WRC statistics for each area of R&D in 2004 (in US million dollar)

Research Area	No. of current projects (sample size)	Average investment in US million dollar per project	Average project length (years)
Water resource management	48	0.164	2.42
Water-linked ecosystems	38	0.123	2.16
Water use and waste management	150	0.079	2,35
Water utilisation in agriculture	38	0.186	3.63

Source: WRC unpublished statistics

Table 12.2 Relative annual investment in research projects in each area of R&D based on the WRC research portfolio (in US million dollar)

	Annual investment in research projects by the WRC				
Research area	03/04	04/05	05/06[a]	Average	%
Water resource management	2,872	3,040	3,131	3,015	31.9
Water-linked ecosystems	1,247	0,973	1,419	2,700	14.3
Water use and waste management	3,297	3,716	3,693	3,569	37.8
Water utilisation in agriculture	1,364	1,494	1,674	1,674	16.0
Total	8,781	9,224	9,918	10,794	100

[a] Budget values.

by professional consultants (more than 20%) and science councils (about 15%) (Table 12.3).

Table 12.4 shows estimates of total expenditure in South Africa on water-centred R&D by the WRC and non-WRC institutions. Non-WRC estimates were obtained from a survey of main providers of water-centred R&D in South Africa. To allow for under-sampling, the sampled non-WRC R&D expenditure was inflated by 50%, before adding the WRC expenditure, to arrive at estimates of gross national expenditure on water-centred R&D. Non-WRC expenditure follow the WRC investment patterns, giving highest shares to WU&WM and WRM. WRC sponsors about 60% of all water-centred R&D investment expenditures in the country.

Table 12.5 presents information on knowledge and capacity-related outputs of WRC research projects completed over a period of 1 year. Assuming that a similar number of projects come to an end annually, these figures may be taken as typical annual outputs of investments in water research. On average, a project yields 0.84 Masters Degrees, 0.27 Ph.D. degrees, 1.43 refereed publications and 3.45 conference presentations.

Table 12.3 Contributions of various categories of research providers to water-centred R&D in South Africa, based on WRC statistics in 2004, in number of projects (N) and US million dollar per annum

	Category of research provider									
	University/ technikon		Science council		Professional consultant		Water/waste utility		NGO	
Research Area	N	US million dollar per year	N	US million dollar per year	N	US million dollar per year	N	US million dollar per year	N	US million dollar per year
Water resource management	27	1,75	11	0,72	8	0,76	1	0,03	1	0,02
Water-linked ecosystems	23	1,59	6	0,27	7	0,27	1	0,02	1	0,04
Water use and waste management	81	2,74	9	0,3	37	1,38	16	0,43	7	0,16
Water utilisation in agriculture	20	0.99	13	0,64	4	0,23	1	0,1	0	-
Total	151	7,07	39	1,93	56	2,64	19	0,58	9	0,22
%	55.1	56.8	14.2	15.5	20.4	21.2	6.9	4.7	3.3	1.8

Source: WRC unpublished statistics.

Table 12.4 Estimated total South African expenditure on water-centred R&D (2004 data in US million dollar)

Research area	National average annual expenditure on water-centred R&D (kR)					
	WRC-supported R&D					
	WRC-supported projects[a]	Total WRC R&D expenditure[b]	Non-WRC R&D expenditure (as sampled)	Non-WRC R&D expenditure – adjusted to compensate for under-sampling[c]	Total national R&D expenditure	% of total investment
Water resource management	3.015	3.869	1.626	2.439	6.309	29.2
Water-linked ecosystems	1.348	1.861	1.098	1.646	3.509	16.3
Water use and waste management	3.569	4.749	2.076	3.114	7.862	36.4
Water utilisation in agriculture	1.511	1.970	1.287	1.931	3.901	18.1
Total	–	12.448	–	9.130	21.580	–
%	–	57.7	–	42.3		100

[a] Estimate based on 3-year average centred on 2004.
[b] Includes expenditure on other R&D support, knowledge dissemination and research portfolio management and administration.
[c] 50% added to sampled expenditure.

Table 12.5 Knowledge dissemination and capacity-building resulting from projects ending over a period of 1 year (based on WRC statistics from 1 April 2003 to 31 March 2004)

Research area	Number of projects completed	Project outputs			
		Masters degrees	Ph.D. degrees	Refereed publications	Conference presentations
Water resource management	37	43	11	102	177
Water-linked ecosystems	16	15	3	21	52
Water use and Waste management	54	34	12	41	113
Water utilisation in agriculture	15	11	7	11	79
Total	122	103	33	175	421

12.5 Impact of Research

To assess the impacts of WRC funded research projects, we report summaries of the results of a study commissioned by the WRC for this purpose. The said study employed cost benefit analysis (CBA) to assess the impact of the following sample of WRC funded projects (WRC 2004): Surface Water Resources of South Africa, Biological Nutrient removal and Combined Services Model. In the following sections we summarise the economic contribution of each project separately.

12.5.1 Surface Water Resources of South Africa

It is difficult to imagine how water resources development in South Africa would have taken place without the pioneering studies of the Hydrological Research Unit (HRU) of University of the Witwatersrand. In 1982, these studies led to a six-volume report, presenting guidelines for estimating water resources in South Africa. Hydrologists and planners relied heavily on these guidelines for water resource analysis, planning and management. The first WRC funded studies were initiated in the early 1970s. The most prominent and long lasting development was probably the well-known Pitman model, which has become the de facto planning model for most of South Africa's bulk water development. After a series of workshops in 1985, a survey known as the "Surface Water Resources 1990 (WR90) was planned. In a 5-year project, a consortium of engineers reassessed South Africa's water resources.

In 1995, a series of reports, consisting of one user manual, six volumes of data, and six books of maps were produced. The data has also been made available via the World Wide Web. Judging by the rate at which these products are being requested by consultants and other interested parties, their impact will be as far-reaching as that of the original set of reports.

The aim of analysing the Surface Water Resources project was, therefore, to compare the costs of funding the research activities of the project with the benefits accruing from this research. It was accepted that most of the benefits would be realised in assisting the planning process. The benefits would result from improved accuracy, and cost savings arising from using the outputs of the project. The costs for the research of the Surface Water Resources project consisted of two elements. The first component is the direct cost of the research and, the second is the administrative cost incurred by the WRC, giving a total cost of US $0.8 million. As mentioned above, the main benefit from this research is the time saved by professionals in using the project output results instead of carrying out primary research on their own to obtain the same data and information. To determine the difference in time usage, a number of case studies were identified in consultation with civil engineering firms and were analysed to determine an average time saving factor. The case studies and other sources indicate that, if the relevant information and data were not available, it would have taken between four and six times longer to undertake and complete the studies to generate the needed information.

A survey was conducted, amongst four engineering firms, to determine the extent of the usage of WR90 and the budgets involved. The survey estimated an average time spent on hydrology analysis of a project to be 80 h per project inwabsence of the WR90 output. Each firm performed an average of 16 projects per year, with 80 h per project being used (at US $54 per h) to calculate the usage costs by the four firms. It is estimated that the four firms surveyed command about 40% share of the market in this field. The CBA estimated a net present value (NPV) of US $3.8 million for this project, which is about five times actual project costs[1].

12.5.2 *Biological Nutrient Removal (BNR)*

The development of the capability to remove nutrients (Nitrogen and Phosphorous) from municipal waste waters became critically important when the rivers draining the industrial and commercial centres in the Gauteng area became highly eutrophic. The WRC considered this a priority research area for an extended period. Although the WRC has funded research in both chemical and biological nutrient removal, the emphasis of the impact assessment has been on biological nutrient removal, as chemical removal contributes to the salinisation of receiving waters. Research into biological nutrient and phosphate removal was first supported by the WRC in 1973, very soon after Dr. JL Barnard first found that phosphorous could be removed biologically. Since that time, research was funded at various institutions, initially with the CSIR, and subsequently, with the Department of Civil Engineering of the University of Cape Town and the Johannesburg City Council playing leading roles in both the development of the process and its understanding. This research led to the development of a number of different process configurations. Some 57 new generation activated sludge plants are operating in the country. This work also led to a number of publications, such as the Theory and Design of Nutrient Removal

[1] See WRC (2004) for further details.

Activated Sludge Processes (Ekama et al 1984), for which there has been widespread international demand. As a result of their research work, researchers from the University of Cape Town formed part of the team that developed the Activated Sludge Models No 1 and 2 of the International Association on Water Quality (IAWQ) for nutrient removal, which has become the internationally accepted model for biological nutrient removal.

For purposes of the CBA, it was necessary to differentiate between the two processes: the historical chemical method, and the new BNR method for the treatment of waste water. Both processes can result in water of a similar quality. However, the chemical method entails the addition of chemicals while the BNR process is more capital intensive. Consequently, it was the objective of this CBA study to compare the capital and operational costs of the two processes. In addition, possible (additional) benefits of the BNR process were also taken into account.

Secondly, the question was whether any positive difference between the cost ratios of the two technologies (which constitutes the benefits of the BNR project) is more than the cost of the relevant WRC research. However, it was noted that it was not only the WRC's contribution which made the BNR process technically viable. This argument was taken into account in the CBA calculations and discussions were held with operators of BNR plants and civil engineers in Gauteng who specialise in designing these plants. Non-availability of accurate cost data complicated the comparative studies. As a result of these discussions, it was, therefore, decided to develop a scenario in which a BNR plant could be compared with a chemical plant of the same capacity by means of a CBA approach. For this purpose, a plant capacity of ten million litres of water per day was selected. The cost element in the CBA is defined as the cost of the WRC research relating to the BNR process. For purposes of the CBA analysis, a total research cost of US $62,500, at 1999 prices, was considered. The benefits of this project were defined as the difference between the costs of the chemical process, and those of the BNR process, to achieve the same relative quality of purified water. The CBA analysis took into account the total BNR capacity of the additional 375 BNR plants that were constructed since 1984. It was estimated (no accurate figures were available) that the 375 plants have a total daily capacity of 65,625 million litres of water.

It is accepted that this total capacity of 65,625 million litres of water is not 100% attributable to the relevant WRC supported BNR research. The study used a conservative assumption that 25% is in fact a direct result of the relevant WRC research. Total benefits of US $11 million have been estimated under this scenario which are considerably higher than the research cost incurred by the WRC[2]. If one accepts that 50% of the total BNR capacity is a direct result of the relevant WRC research, the benefits are even higher (i.e. US $22 million).

The CBA calculated a NPV of this research project of US $22 million, which is several times the WRC cost of about US $62,000.

[2] See WRC (2004) for further details.

12.5.3 *The Combined Services Model (CSM)*

One of the aims of the WRC is to supprt the supply of clean drinking water to everyone. This caused the WRC to become involved, in 1992, in the "Water and Sanitation in Urban areas: Financial and Institutional Review", in partnership with the Development Bank of Southern Africa (DBSA). At the start of the project it was only envisaged to develop a water supply model. This was later adapted for sanitation services with the assistance of the Durban Corporation. The model used for water supply and sanitation was also extended to include roads, stormwater drainage, electricity and solid waste management. This task resulted in the creation of a number of spreadsheets that were, however, not fully integrated nor easy to use. On the other hand, this modelling system provided useful information in support of decision-making.

Recognising that there was a very real need for a user-friendly model that would assist with long term infrastructure investment planning, the DBSA proposed a co-funding arrangement with the WRC to produce a Combined Services Model (CSM). The Palmer Development Group was commissioned to undertake the work during 1996. The outcome of this project was that the CSM, a sophisticated computer-based model, was developed as a predictive model of the expenditures and cash flows associated with the provision of municipal infrastructure services for residential consumers. The main advantage of the model is that it contributes to the assessment of appropriate services in line with what consumers can afford. To assess the viability of a planned infrastructure, the combined services model calculates the capital and operating costs and borrowing requirements associated with the investment programme in both nominal and real terms; and assesses the affordability of the programme to households and the local authorities.

The WRC funded part of the relevant research to the extent of US $58,000. A CBA was commissioned to test the hypotheses that, by using the CSM, the local authority would save money in the planning process; secondly, that by using the CSM, the local authority would save the ratepayers money since ratepayers would receive a service commensurate with their needs and ability to pay. The benefits calculated for the ratepayers and for the local authorities could not be added together since they are mutually exclusive, and either one or the other could be accepted as the benefits emanating from the application of the CSM model.

12.5.3.1 Local Authority Savings

Below are present values depicting differences in costs, if 72 local authorities either make use, or do not make use of the CSM as a planning tool (Table 12.6).

Table 12.6 Value of costs with and without CSM

	US dollar
Present value of costs without CSM	1,298,922
Present value of costs with CSM	144,983
Present value of CSM benefits	1,153,939

1999 prices.

Table 12.7 CBA results of the CSM

1999 Economic prices	Local authority savings Present Value 8% US million dollar	Savings for rate payers Present value 8% US million dollar
Cost		
Cost of WRC Research Projects	–0.054	–0.054
Benefits	1.15	18.59
Net present value	1.09	18.54

12.5.3.2 Savings for Rate Payers

In calculating the benefit to ratepayers, the existing situation of local authorities was compared with the end result, after the implementation of the proposed strategy of using the CSM. It was assumed that only 10% of local authorities make use of the CSM. Based on this assumption, a NPV of the savings of US $1.8 million has been calculated. The CSM was also nominated by the DBSA for a National Productivity Award, which it eventually received. The model is actively promoted by the DBSA and they have also made use of the model in the planning process as a condition before any loans for infrastructure projects are approved. Research has also shown that nearly 100 local authorities have already made use of the model, with some authorities using it more than once. In addition, the CSM is also used by private consultants in planning processes.

The CBA results of the CSM are given in Table 12.7.

As with all organisations, it is necessary, from time to time, to evaluate the performance of the WRC. This study was focused on the evaluation of the economic contribution of WRC research, which only covered one element of an array of criteria that have to be applied to determine the overall efficiency of its operations, in terms of the set objectives. Using the CBA techniques applied to the projects identified for this purpose, it was established, with a reasonable amount of certainty that, in terms of NPV estimates, the WRC's research efforts have made significant contributions towards all its focus areas, including; technology transfer, knowledge empowerment, determination of research priorities and directing research into priority areas.

12.6 Future Needs and Challenges

The South African water sector currently appears to be well-supported, and has received substantial R&D investments. Having a dedicated body, such as the WRC, to lead such support is, undoubtedly, the most important factor behind this healthy situation. The WRC ensures that a strategic view on R&D needs is maintained and that the most critical R&D needs receive timely attention and, most importantly, that the water-centred knowledge base is continuously renewed and maintained.

The contribution of the WRC to the R&D leadership and gross national expenditure on R&D in South Africa facilitates, up to a point, the gathering and provision of national statistics on water-centred R&D. Many of these have been translated into R&D input and output indicators which provide a baseline for establishing future trends. However, there remains great uncertainty attached to statistics pertaining to water R&D conducted independently of the WRC. There is a definite need to establish a system, whereby more precise estimates can be obtained, of contributions to the national effort by role-players other than the WRC. Putting such a system in place presents a challenge which, if successfully met, would allow for better reporting in future and also a greater degree of refinement in strategic planning of R&D in support of the South African water sector.

Acknowledgements The information presented in this paper emanates, in part, from research work undertaken by Dr George Green (2005) on behalf of the WRC and from the WRC Cost Benefit Analysis Study (WRC 2004).

References

DWAF (2005) The national water resource strategy for South Africa. Department of Water Affairs & Forestry, 2005. Government Printer, Pretoria

Ekama GA, Marais GVR, Siebritz IP, Pitman AR, Keay GFP, Buchan L, Gerber A, Smollen M (1984) Theory, design and operation of nutrient removal activated sludge processes. Water Research Commission Report, WRC TT 16/84, Pretoria

Green G (2005) Water research and development. Water research commission (unpublished report), Pretoria

RSA (1970) Report of the commission of enquiry into water matters. R.P. 34/1970. Republic of South Africa, Pretoria

RSA (1971) The Water Research Act – No. 34 of 1971. Republic of South Africa, Pretoria

RSA (1997) Water Services Act. Act 108 of 1997. Government Gazette No. 18522. Office of the President, Cape Town

RSA (1998) National water act. Government gazette, vol 398. Office of the President, Cape Town, 26 August 1998

WRC (2004) Lessons learned from cost benefit of selected WRC research projects. Water Research Commission, WRC TT 226/04, Pretoria

Chapter 13
Transboundary Water Management Issues Under the NWA and Regional Collaboration, Policies and Conventions

Reginald Tekateka

Abstract In order to contextualise the current South African approach to transboundary water resources management this chapter provides a brief background of the political and legal developments leading towards the establishment of the Orange/Senqu River Commission and the ongoing activities of the Commission. This illustrates the change of direction of South Africa's approach to transboundary water management after 1994.

Thereafter the chapter provides an overview of the current SADC policy framework, namely the Regional Water Policy and Regional Water Strategy and the SADC Protocol on Shared Watercourses as the most important regional instruments governing transboundary water resources management. The increasingly important developments at the continental level, particularly under the umbrella of the African Ministers' Council on Water (AMCOW), are also illustrated in this section.

Finally, the chapter describes the Incomaputo-Agreement,[1] to which South Africa is a Party, as an advanced model of transboundary cooperation in water management. The Incomaputo-Agreement is the first basin-wide water management agreement concluded in the southern African region and covers a comprehensive range of issues including water allocation between countries, water quality control and emergency measures. Using the Incomaputo-Agreement as an example, the chapter discusses issues of assessing water entitlements as well as institutional arrangements to implement a basin-wide agreement and enforce its provisions domestically in South Africa.

---

[1]Tripartite interim agreement between the Republic of Mozambique and the Republic of South Africa and the kingdom of Swaziland for co-operation on the protection and sustainable utilisation of the water resources of the Incomati and Maputo watercourses (Republic of Mozambique, Kingdom of Swaziland & Republic of South Africa 2002).

R. Tekateka (✉)
Chair: African Network of River and Lake Basin Organisations (ANBO)
e-mail: tekatekar2@gmail.com

B. Schreiner and R. Hassan (eds.), *Transforming Water Management in South Africa: Designing and Implementing a New Policy Framework*, Global Issues in Water Policy 2,
DOI 10.1007/978-90-481-9367-7_13, © Springer Science+Business Media B.V. 2011

Keywords SADC Regional Water Policy • SADC Regional Water Strategy • SADC Revised Protocol on Shared Watercourses • Institutional arrangements • Water allocation • Basin-wide agreement • Inco-Maputo agreement • ORASECOM • Transboundary co-operation

13.1 Introduction

"The purpose of this Act is to ensure that the nation's water resources are protected, used, developed, conserved, managed and controlled in ways which take into account amongst other factors -…(i) meeting international obligations;…" Section 2 (i) of the National Water Act 36 of 1998 (RSA 1998).

South Africa shares four rivers with its six neighbours: the Orange-Senqu River is shared with Botswana, Lesotho and Namibia, the Limpopo River with Botswana, Mozambique and Zimbabwe, and the Incomati River and the Maputo River are shared with Mozambique and Swaziland.

South Africa's approach to transboundary water resource management has substantially changed after the country's international isolation was overcome with its transformation to democracy. Although there has previously been cooperation on shared waters between South Africa and its neighbours, it is only now, after the democratic South Africa came into being, that the country's policies on transboundary water resources are solidly embedded in a regional framework and strongly guided by regional policies and international law.

Fig. 13.1 Shared river basins of South Africa (Source: DWA)

Today, South Africa is a member of the Southern African Development Community (SADC) and subscribes to the SADC Regional Water Policy and SADC Regional Water Strategy. Likewise, South Africa has ratified and is bound by the Revised SADC Protocol on Shared Watercourses (hereafter the SADC Protocol), the overarching legal instrument governing transboundary water resources management in the southern African region. South Africa is also one of the few states that have ratified the 1997 UN Convention on the Law of the Non-Navigational Uses of International Watercourses (hereafter UN Convention).[2]

The shift in approach towards one in line with broader regional integration is also reflected in South Africa's domestic water law. Whereas the previous Water Act 54 of 1956 did not refer to international obligations, meeting international obligations is now explicitly listed as a purpose of the new National Water Act 36 of 1998 and is second only to the Reserve, which is granted as the only "right" in the Act.

In order to contextualise the current South African approach to transboundary water resources management this chapter will provide a brief background of the political and legal developments leading towards the establishment of the Orange/Senqu River Commission as well as the ongoing activities of the Commission. This will illustrate the change of direction South Africa's approach to transboundary water management has seen after 1994.

Thereafter the chapter provides an overview of the current SADC policy framework, namely the Regional Water Policy and Regional Water Strategy and the SADC Protocol on Shared Watercourses as the most important regional instruments governing transboundary water resources management. The increasingly important developments at the continental level, particularly under the umbrella of the African Ministers' Council on Water (AMCOW), will also be illustrated in this section.

Finally, the chapter describes the Incomaputo-Agreement,[3] to which South Africa is a Party, as an advanced model of transboundary cooperation in water management. The Incomaputo-Agreement is the first basin-wide water management agreement concluded in the southern African region and covers a comprehensive range of issues including water allocation between countries, water quality control and emergency measures. Using the Incomaputo-Agreement as an example the chapter will discuss issues of assessing water entitlements as well as institutional arrangements to implement a basin-wide agreement and enforce its provisions domestically in South Africa.

---

[2]Since the UN Convention has so far not received the required number of 35 ratifications, it is yet to come into force. Nevertheless the UN Convention is considered to be an important instrument of international water law and it is widely accepted that its key principles form part of the body of international customary law (UN 1997).

[3]Tripartite interim agreement between the Republic of Mozambique and the Republic of South Africa and the kingdom of Swaziland for co-operation on the protection and sustainable utilisation of the water resources of the Incomati and Maputo watercourses

13.2 Changing Approaches to Transboundary Water Resources Management – the Orange Senqu River Basin Example

The largest basin South Africa shares with its neighbours, the Orange Senqu River basin, is at the same time the economically most important one for the country. Much of the water for Gauteng Province, South Africa's economic heartland, is provided from the Orange Senqu River and a significant amount is supplied through the Lesotho Highlands Water Project (Earle et al. 2005).

The institutional and legal developments related to the management of the Orange Senqu River Basin provide an illustrative example of how South Africa's approach to the management of its shared rivers has changed over time. There has been long-standing cooperation over the water of the Orange Senqu River. The nature of cooperation, however, has changed significantly in more recent years. For decades cooperation was restricted to the bilateral level and the first basin wide agreement, the agreement establishing the Orange Senqu River Commission (ORASECOM), was only signed in 2000.

13.2.1 From Bilateral Cooperation to the Establishment of ORASECOM

Initial cooperation on matters relating to the Orange Senqu River Basin occurred from 1948 under the auspices of the standing committee on water of the Southern African Regional Commission for the Conservation and Utilisation of the Soil (SARCCUS) (Turton 2004). After a relatively long period without further institutional and legal developments, a new era of bilateral cooperation began in 1978 when South Africa and Lesotho established a Joint Technical Committee (JTC) to investigate the feasibility of the Lesotho Highlands Water Project (Heyns 1995 quoted in Turton 2004). The Lesotho Highlands Water Project is the largest international interbasin transfer in the world (DWAF 2004). It delivers water from the upper reaches of the Orange Senqu River in Lesotho to South Africa in exchange for royalties paid by South Africa to Lesotho.

As the parties subsequently decided to go ahead with the project, the Lesotho Highlands Water Project Treaty (LHWP-Treaty)[4] was signed in 1986 and covers in detail aspects of the design, construction, operation and maintenance as well as the institutional arrangements needed to manage the project (Earle et al. 2005). At the organisational level the treaty established two autonomous statutory bodies. The Lesotho Highlands Development Authority (LHDA) is responsible for the manage-

[4]Treaty on the Lesotho Highlands Water Project Between the Government of the Republic of South Africa and the Government of the Kingdom of Lesotho (1986).

ment of the dam construction and related issues in Lesotho, and the Trans-Caledon Tunnel Authority (TCTA) is responsible for the management of the delivery tunnels into South Africa (Turton 2004). Furthermore, the treaty created the Joint Permanent Technical Commission (JPTC) which consists of delegates from both riparian states and was responsible for coordinating the two parastatals and for reporting back to the respective governments (Turton 2004). In 1999 the JPTC was upgraded into the Lesotho Highlands Water Commission (LHWC) by Protocol VI to the LHWP-Treaty.

The gradually intensifying cooperation between Lesotho and South Africa before, during and after the construction of the Lesotho Highlands Water Project was driven by both countries' interest in the completion of the project, hence did not at that time envisage the establishment of a more comprehensive management regime for the basin. Consequently, the treaty deals solely with matters relating to the construction and maintenance of the project and the water delivery to South Africa.

Although arguably not being fully in line with the key principles of international water law, as, for example, expressed in the SADC Protocol, the LHWP-Treaty continues to be of importance for the management of the basin, since the rights and obligations established in it are not affected by the subsequent ORASECOM-Agreement (see more detailed discussion below).

The establishment of ORASECOM only materialised after a further period of bilateral cooperation, this time between South Africa and Namibia. After Namibia became independent from South Africa in 1990, the two countries in 1992 signed an agreement establishing a Permanent Water Commission (PWC), which replaced a Joint Technical Committee that was established in 1987 and functioned in the transitional period before Namibia's independence (Heyns 2004). The PWC was established to advise the Governments on matters pertaining to the development of the Lower Orange River where it forms the border between South Africa and Namibia (Heyns 2004). At the same time South Africa and Namibia signed an agreement on the establishment of a Joint Irrigation Authority (JIA) to implement the Vioolsdrift and Noordoewer Joint Irrigation Schemes (Turton 2004).

Until the formation of ORASECOM, cooperation occurred primarily on the bilateral level. The negotiations to form a basin-wide commission started in the wake of the transition to democracy in South Africa and parallel to events occurring at regional level. As described by Heyns (2004) South Africa hosted a conference of SADC Ministers responsible for water resources management in 1995, which among other things dealt with infrastructure development and the management of water issues on shared rivers in the SADC region. The proposal to create a SADC Water Sector was supported by all Orange Senqu River basin states and the regional developments prompted them to discuss the possibility to create a multilateral basin commission (Heyns 2004). The subsequent process of negotiations was finally concluded when ORASECOM was formally established on 3 November 2000.

Whereas the cooperation between Orange Senqu River basin states has moved from an era of bilateral cooperation to a basin-wide approach, the bilateral agreements and organisations set up prior to the formation of ORASECOM continue to exist. Article 1 (4) of the ORASECOM-Agreement states that "existing Commissions will liaise with this Commission in terms of this Agreement".

Equally, the rights and obligations established in prior agreements remain unaffected since Article 1 (3) of the ORASECOM-agreement states that "in the absence of an agreement to the contrary, nothing in this Agreement shall affect the rights and obligations of a Party arising from other agreements in force prior to the date this Agreement comes into force for such a Party". Arguably the most significant aspect of Article 1 (3) is that the (bilaterally agreed) water delivery from Lesotho to South Africa as determined by the LHWP-Treaty, remains unchanged and needs to be taken into account for possible future basin-wide agreements on water allocation. In other words when ORASECOM advises the parties on how to share the Orange Senqu River resources in an equitable and reasonable manner, they have to strike the balance with a certain status quo already established.

Nonetheless, the new management framework for the Orange Senqu River Basin is illustrative of the move away from a bilateral towards a basin-wide management approach, as is the case with other basins which South Africa shares with its neighbours (e.g. Incomati, Limpopo).

13.2.2 Functions and Activities of ORASECOM

As per the ORASECOM-Agreement[5] (Article 1 (2)) the Commission is an international organisation with international and national (in each member state) legal personality. The objectives of the ORASECOM Council (the highest decision making body) are stipulated in Article 4 of the ORASECOM-Agreement, which states that "the Council shall serve as technical advisor to the Parties on matters relating to the development, utilisation and conservation of the water resources in the River System and shall perform such other functions pertaining to the development and utilisation of water resources as the Parties may agree to assign to the Commission".

The objective as outlined in Article 4 of the agreement gives ORASECOM a wide-ranging mandate to discuss issues related to the management of the Orange Senqu River resources and advise the member states on these issues. Furthermore, Article 5 of the ORASECOM-Agreement singles out a number of areas where the commission is requested to take the required measures necessary for advising the parties. These issues are the long-term yield determination, equitable and reasonable utilisation, studies with regard to the development of the resources, stakeholder

[5]Agreement Between the Governments of the Republic of Botswana, the Kingdom of Lesotho, the Republic of Namibia and the Republic of South Africa on the Establishment of the Orange-Senqu River Commission (Republic of Botswana, Kingdom of Lesotho, Republic of Namibia & Republic of South Africa 2000)

involvement, data collection and sharing, pollution prevention, measures for emergency situations, information exchange and consultation between parties and measures for the prevention and settlement of potential disputes.

In practise, ORASECOM has focused on a multitude of projects, largely with the support of international donor agencies, such as GtZ, EU, French GEF, UNDP/GEF and InWEnt. Driven by the ORASECOM Secretariat the various support programmes are currently being coordinated into one programme framework for the basin to avoid overlaps and create synergies between the activities supporting the basin. This coordination effort includes that the reporting requirements of technical advisors brought in by the various initiatives are geared towards and accountable to the ORASECOM Secretariat.

The programme coordination will require strong organisational capacity of the Commission Secretariat in order to be able to play its role in the implementation of the various projects falling under the programme. This capacity will be built in the ORASECOM Secretariat hosted in South Africa. ORASECOM Member States have for a while been working towards the establishment of a Secretariat for the Commission, which will provide administrative support to the Commission and run its day-to-day affairs. Initially an interim Secretariat was established, which has been replaced by a permanent Secretariat headed by an Executive Secretary in October 2007.

A key component of many of the projects to be implemented in the basin will be stakeholder participation. To this end the Commission had started, following an instruction of the water Ministers of the four basin states in 2005, to develop a mechanism for cooperating with stakeholders in the management of the basin. The "Roadmap towards stakeholder participation" has been finalised in April 2007 and will form the basis for stakeholder participation activities in the basin. Importantly, the "Roadmap" is believed to be the first one on the African continent that has been developed through an interactive, participative process between the Commission as well as stakeholder participation experts from southern, East and West Africa (as well as from the US). This process ensured ownership of the "Roadmap" by the Commission and could be an interesting model for the proposed collaboration between governments and civil societies as driven by the AMCOW (see Section 13.3).

The "Roadmap" emphasises the importance of progressive, gradually increasing involvement of stakeholders in the planning and decision-making processes relating to the management of the resources of the basin with the goal of co-management between responsible government organisations and stakeholders. The proposed organisational structures for stakeholder participation activities have been aligned with the ones designed for the TDA-SAP, thereby creating synergies between the two processes that are aimed at ensuring the building of sustainable implementation capacity at the different levels (governments and stakeholders). Both the process of developing the roadmap, and the creation of synergies between activities and alignment of organisational structures could be an interesting model to draw experiences from, as other basins or regions develop their own mechanisms for stakeholder participation.

13.3 Regional Frameworks and Continental Developments

13.3.1 The SADC Framework for Transboundary Water Resources Management

The South African support for this new approach to transboundary water resources management in the Orange Senqu River and other basins has been made possible by the country's democratic transformation. The country's new leadership promoted a development trajectory in which regional integration would play an important role. The development of the country's new water policy and water law was no exception to this approach.

Section 6.9 on International Cooperation of the White Paper on a National Water Policy for South Africa states that "the most important consideration in South Africa's relation with those of its neighbours with which it shares rivers is that there must be respect for each Country's equitable right to water from the shared resource" (RSA 1996:6.9.1). The White Paper continues in expressing South African support for the development of international law to guide the management of shared river systems on an equitable basis and the promotion of the (then applicable) SADC Protocol on Shared Water Course Systems (RSA 1996). It thus crystallised at this stage that South Africa's approach to managing its transboundary waters was to be guided by the principles of international law and integrated with the SADC framework. Consequently South Africa is one of the relatively few countries that have ratified the UN Convention on the Law of the Non-navigational Uses of International Watercourses. Additionally South Africa has, like most other SADC states, ratified the Revised SADC Protocol and subscribes to and actively supports the SADC Regional Water Policy and Regional Water Strategy.

13.3.2 SADC Protocol

From a legal perspective, the SADC Protocol is the key instrument for transboundary water management in the SADC region. Whereas the SADC Regional Water Policy and Regional Water Strategy are important guideline documents, the SADC Protocol establishes a legally binding framework for transboundary water management in the region.

It is important to take cognisance of the fact that the SADC Protocol is a framework agreement. It contains the generic rules for the management of shared rivers within the SADC region, but does not contain basin-specific rules. The latter are to be included in (basin-wide) watercourse agreements, such as the Incomaputo Agreement. The link between the SADC Protocol and the basin-specific rules is made in Article 6 (3) of the SADC Protocol, which states that "watercourse states may enter into agreements, which apply the provisions of this Protocol to the characteristics and uses of a particular shared watercourse or part thereof". The SADC Protocol thus provides the general direction and principles for any future watercourse agreements concluded in the SADC region, while at the same time allowing for the consideration of certain characteristics that may be specific to the watercourse in question (Ashton et al. 2006).

The generic provisions of the SADC Protocol are drafted in line with the provisions of the UN Convention, thus reflecting contemporary international water law. With signing (and ratifying) the SADC Protocol the SADC Member States have expressly undertaken to adhere to the rules of international water law shared by most states in the world relating to the utilisation and management of the resources of shared watercourses (Article 3 (3)). In line with these rules of international water law the SADC Protocol contains the principles of "equitable and reasonable utilisation" (Article 3 (7)) and the "duty to prevent significant harm" (Article 3 (8)). It furthermore, among others, contains provisions dealing with notification and consultation requirements regarding planned measures and rules on pollution prevention, reduction and control.

Equally important to setting out the above-mentioned general principles is that the SADC Protocol establishes an institutional framework at the regional level for the implementation of the instrument. In Article 5 it establishes the SADC Water Sector Organs and mandates them as well as Shared Watercourse Institutions with the implementation of the Protocol.

In practise, the SADC institutions are currently mandated primarily with monitoring functions concerning the application of the SADC Protocol as well as with facilitating the harmonisation of water law and policy between SADC member states. SADC institutions are not mandated with the implementation and enforcement of basin-wide agreements. Where those have been concluded this is done by Shared Watercourse Institutions as well as the domestic institutions in the countries that are party to the basin-wide agreement (see section on Incomaputo-Agreement below).

13.3.3 Regional Water Policy and Regional Water Strategy

The main water related policy documents at SADC level are the Regional Water Policy (RWP) and the Regional Water Strategy (RWS). They lay down regionally agreed policy guidelines concerning water resources management, covering a wide range of topics from infrastructure development, information exchange, capacity building to gender aspects and stakeholder involvement. Although, being policy documents they are not legally binding on the SADC member states; they are important guides for the ongoing harmonisation of national water policies of the SADC member states, many of which are still based on the water policies and laws developed under colonial rule. They also inform the implementation of the SADC Regional Indicative Strategic Development Plan (RISDP) and the SADC Regional Strategic Action Plan (RSAP), the water related development plans of the SADC.

The RWP and the RWS subscribe to the overarching principle of Integrated Water Resources Management (IWRM), defined by the Global Water Partnership (GWP) as "a process which promotes the coordinated development and management of water, land and related resources in order to maximise the resultant economic and social welfare in an equitable manner without compromising the sustainability of vital eco-systems" (GWP 2005). IWRM takes cognisance of the interrelationship between different natural resources and aims at integrating them into a holistic management system (GWP 2005). As such, it is not an end in itself but a means of achieving three key strategic objectives.

- Efficiency to make water resources go as far as possible
- Equity, in the allocation of water across different social and economic groups
- Environmental sustainability, to protect the water resources base and associated eco-systems (GWP 2005)

Key elements of an IWRM based water resources governance framework would, among others, be open and transparent institutions, inclusive and participative decision-making, equitable access to the resource and coherent and integrated policies (GWP 2005).

The RWP and RWS contain these elements, and furthermore, recognise the importance of regional cooperation over water resources and the need to manage water resources in an integrated manner (Malzbender and Earle 2007), specifically highlighting the need for regional integration (Policy 3.1) as well cooperation between all affected (water use) sectors (Policy 3.3) (SADC 2005).

The RWP promotes the establishment and development of transparent institutions and the involvement of stakeholders in water management decision-making. In line with the provisions of the SADC Protocol the policy calls for the establishment of Shared Watercourse Institutions (SWCI) on each shared watercourse (Policy 9.2.2), which shall promote stakeholder participation in decision-making (Policy 9.2.8) (Malzbender and Earle 2007). In Chapter 10 the policy deals exclusively with stakeholder participation and capacity building, stating that water resources management and development at all levels shall be based on a participatory approach (Policy 10.1) and that stakeholders need to be empowered to effectively participate in such decision-making (10.1.2) (Malzbender and Earle 2007).

This policy framework is increasingly being reflected at national level in SADC states. Some countries in the region have already revised their water laws and policies and in this process accounted for the integration of IWRM principles in their national policy and legal framework. Other countries are still in the process of revising their policies and laws and while doing so are guided by the principles enshrined in the SADC instruments (Malzbender and Earle 2007).

Importantly, the fact the SADC RWP and RWS were co-developed by stakeholders through a participative process should give them a strong persuasive voice in guiding policy harmonisation between member states. Against that background, the ongoing harmonisation of policies and laws in and between Member States and with SADC policies are likely to ensure a coherent regional water resources management framework that takes the important principles of IWRM into account throughout the region. The RWP and RWS are key elements of the implementation phase of the SADC Protocol. The advances made in the SADC region could become reflected continentally through the efforts of AMCOW. As one of the key drivers of the continental processes, South Africa recommends the SADC advances as a potential model for an African approach.

13.4 Developments at Continental Level

In addition to the regional policies at SADC level, there is increasingly a drive for policy development and joint decision making at the continental level. The Africa Water Vision is one such example, the various initiatives being discussed under the umbrella of AMCOW and NEPAD being others. Whereas regional integration in the SADC region is more advanced than in other parts of Africa, cooperation on issues around water management, and making best use of the development potential of water, is generally increasing in Africa.

These initiatives are strongly driven by AMCOW, which aims to facilitate greater cooperation on water resources management between countries and regions as well as between different stakeholder groups. Recently, the African Union (AU) has in principal agreed to recognise AMCOW as a Specialised Technical Committee (STC) of the AU. This strengthens AMCOW's mandate and enhances its ability to respond to challenges, formulate strategies and engage with partners in a coordinated fashion. Given their existing experience with respect to cooperation over transboundary water resources, the SADC states can make a valuable contribution to these processes.

A key component of AMCOW's initiatives is cooperation with stakeholders, which are engaged with AMCOW at various levels. Following decisions taken by AMCOW at its sixth Ordinary Session in May 2007, AMCOW is to establish a multi-Stakeholder Forum, such that:

Shared Watercourse Institutions, organised under the African Network of River Basin Organisations (ANBO), are themselves stakeholders under AMCOW, thus facilitating alignment between political developments at continental level and water management at basin level. ANBO is now officially recognised as a sub-committee of AMCOW.

The new chair of AMCOW was mandated to sign a Memorandum of Understanding (MoU) with the Global Water Partnership (GWP), thus ensuring the latter is more closely engaged in advancing the African water agenda through the involvement of its extensive pool of African professionals and experts in the activities of AMCOW.

The new chair of AMCOW is mandated to also sign a MoU with the Africa Civil Society Network on Water and Sanitation (ANEW), which has been recognised as the official civil society cooperation partner of AMCOW. This partnership will strengthen the voice of civil society with respect to water resource governance on the continent.

It is envisaged that similar cooperation arrangements could be entered into with continental bodies of Parliamentarians, city and local governments, centres of excellence, etc. The broad aim of AMCOW in entering into these arrangements is to benefit from the inputs of all these stakeholders in its own planning and decision making processes.

Particularly, ANBO could potentially play an important role in facilitating interaction between countries at basin level. For example, in cases where basin agreements have been jointly developed and countries are expected to sign (or have signed but not ratified) but cause undue delays without explanation to the other countries, ANBO could potentially play a role as an independent mediator. At the same time, SADC could also develop its own mechanism for mediation for such cases and draw upon ANBO to support any arising mediation exercises if deemed necessary.

In another example, it is important that information on the programme coordination exercise currently being undertaken by ORASECOM is shared with ANBO for lessons learnt to support similar exercises in other African basins in the future. At the same time the ORASECOM example could also be used as a case study for donor coordination to be shared and discussed with donors within the context of the Paris process and SADC's own Windhoek declaration.

A particular focus of AMCOW's efforts to strengthen cooperation between, inter alia, governments and civil society in water resources development and management is the latter's potential contribution in the promotion of sustainable water infrastructure development through the inclusion of stakeholders in decision-making processes. The African Ministers' Conference on Hydropower and Sustainable Development held in Johannesburg in March 2006 was a milestone event in this regard. Not only did the Ministers agree on initiatives to jointly develop Africa's hydropower but in the Declaration placed specific emphasis on the fact that the development of hydropower must be implemented in accordance with national, regional and international agreements and respect the principles of sustainable development, and requires an integrated and holistic approach (AMC 2006). This commitment to cooperation and adherence to legal principles is evidence that African countries, including the SADC member states, place strong emphasis on cooperation over shared water resources (Earle and Malzbender 2007). Importantly, the Ministers' decision emphasised the importance of involving stakeholders at all stages during the planning and implementation of infrastructure projects. This high-level political commitment puts Africa at the forefront of a process that sees civil society involved in infrastructure development and water resources development.

As a direct consequence, an uncommon joint initiative by SADC and the East African Community (EAC) aimed at advancing the implementation of the Ministerial Action Plan and titled the Major Water Infrastructure Development (MWID) Seminar, took place in Swaziland in July 2007. Bringing together over 70 water and other relevant sector professionals from regional governments in East and Southern Africa, as well as representatives of NGO's, the private sector and the major regional financial institutions, the seminar produced a comprehensive set of recommendations, central among which was the need to develop tools to form and build the capacity of the broad range of stakeholders in the area of large scale infrastructure development. It is the hope of the two sub-regions that other sub-regions will also hold similar seminars to develop a truly African approach to the development of large infrastructure projects in Africa, as the continent readies itself to implement NEPAD's infrastructure development programme. It is important that Development Partners and International Finance Institutions acknowledge and strengthen this process.

13.5 Basin-Wide Agreements Within the SADC Framework – the Example of the Incomaputo-Agreement

The Incomaputo-Agreement is a manifestation of South Africa's commitment to transboundary water resources management under the governing rules of international law, particularly the SADC Protocol. With the signing of this agreement the three parties to it (Mozambique, South Africa, Swaziland) have established the first basin-wide agreement in the SADC region, which establishes a comprehensive, basin-wide management regime. Whereas there are other agreements in place that involve all basin states (e.g. the agreements establishing ORASECOM, OKACOM and LIMCOM), they have so far not progressed to the stage where a comprehensive basin management regime is established.

At the same time, the Incomaputo-Agreement is the first regional watercourse agreement to be concluded after the signing of the SADC Protocol (Ashton et al. 2006). While the SADC Protocol was not in force when the Incomaputo-Agreement was concluded (in 2002), it (the SADC Protocol) had already been negotiated and signed by member states. It was clear that the SADC Protocol would come into force in the near future and consequently the Incomaputo-Agreement was drafted following the rules and the spirit of the SADC Protocol. More so, the Incomaputo-Agreement is directly rooted in the SADC Protocol as it is a shared watercourse agreement in terms of Article 6 (3) of the SADC Protocol.

Throughout the agreement text, the Incomaputo-Agreement complies with the principles set out by the SADC Protocol. As required by Article 6 (4) of the Protocol the Incomaputo-Agreement defines the waters to which it applies, namely the Incomati and Maputo Watercourses. In terms of substantive rules, the Incomaputo Agreement contains those rules of international water law, which are today considered to be fundamental principles of watercourse agreements. These rules, which should be included in every international watercourse agreement (Vinogradov et al. 2003) are the principle of "equitable and reasonable utilisation" (Article 3 (b) of the Incomaputo-Agreement) and the "obligation not to cause significant harm" to other watercourse states (Article 6 (3) of the Incomaputo-Agreement). Both rules are contained in the UN Convention (Articles 5 and 7) and the SADC Protocol (Articles 3 (7) (a) and 3 (10) (a)), thus showing that the Incomaputo Agreement was clearly drafted in line with these international water law frameworks.

13.5.1 Water Allocations

The Incomaputo-Agreement contains detailed provisions on flow regimes, which form the basis of water allocations to the riparian countries. In terms of Article 9 (2) of the Incomaputo-Agreement any abstraction of water from the system shall be in accordance with the flow regimes as determined in Annex I of the agreement. The minimum flow into Mozambique that was established by the Piggs Peak agreement and was historically the first water allocation agreement between the parties, has been included in the flow regime of the Incomaputo-Agreement. Yet, the latter one is far more detailed and regulates flow regimes and water allocations in a much more comprehensive manner.

An interesting feature of the Incomaputo-Agreement is that it does not only regulate minimum flows, but with the water allocations to countries also establishes priority uses and allocates specific amounts of water for each category of use (see for example Article 4 of Annex I for the Incomati watercourse) (Ashton et al. 2006). Thereby it distinguishes between first priority supplies (domestic, livestock and industrial use), irrigation supplies and a calculated runoff reduction as a consequence of afforestation and determines the exact amount for each of the uses for every catchment that forms part of the Incomati or Maputo watercourses (Ashton et al 2006).

Furthermore, the agreement takes cognisance of the climatic conditions prevailing in the basin and makes provision for situations where the allocations cannot be met, i.e. during drought conditions. Article 4 (5) of Annex I determines that during droughts, all parties must reduce their water use sequentially in the following order: Water use for irrigation must be reduced first, followed by reductions in first priority uses. A reduction of the water for the riverine and estuarine ecosystems shall only be allowed under extreme drought conditions (Ashton et al. 2006). With this system of water allocations, priority uses and reductions in emergency situations, the Incomaputo Agreement introduces a flexible system, which addresses the specific concerns of a potentially drought-stricken region in an adequate manner (Ashton et al. 2006).

13.5.2 Other Key Areas

In addition to its water allocation and use specification provisions, the Incomaputo Agreement, among others, contains provisions on water quality and pollution prevention, data and information exchange and prior notification and consultation on planned measures.

Rules on water quality are contained in Article 8, which establishes a comprehensive regime for water quality management and pollution prevention, including the development of a classification system by the Tripartite Permanent Technical Committee (TPTC) (Ashton et al. 2006).

Provisions on data and information exchange can be found in Article 12 and Article 13 of the agreement which require parties to adhere to certain procedures concerning information and notification about planned measures, provided the planned measures have the potential of or are likely to cause significant transboundary impact. The specific requirements for information and notification are not stipulated in the agreement. Instead the agreement refers to the requirements set out in Article 4 (1) of the SADC Protocol, making it clear that it needs to be understood as a specific part of the SADC wide transboundary water management framework as provided for by the SADC Protocol (Ashton et al. 2006).

13.5.3 Implementation and Enforcement

The main role in the implementation of the Incomaputo-Agreement on the international level is played by the Tripartite Permanent Technical Committee (TPTC). The TPTC was established by the parties in 1983 and is a Shared Watercourse Institution in terms of Article 5 (3) of the SADC Protocol, thus embedded into the watercourse management framework created by the SADC Protocol.

According to Article 5 (1) of the Incomaputo-Agreement the TPTC shall be the joint body for co-operation between the parties and is mandated with the implementation of the agreement. The Committee's powers are generically defined in Article 5 (2) of the agreement, stating that the TPTC "shall exercise the powers established in this Agreement, as well as those conferred by the Parties in order to pursue the objectives and provisions established herein" (Ashton et. al. 2006). More detailed powers of the TPTC are specified throughout the agreement. In terms of Article 8, the TPTC is responsible for the implementation of the necessary measures to prevent, reduce and control pollution of the watercourse. Furthermore, in terms of Article 12, the TPTC shall be the forum for information and data exchange between the parties and shall determine the procedures for environmental impact assessments related to planned measures involving significant transboundary impact of substantial magnitude (Article 13 (3)) (Ashton et al. 2006).

Substantial powers are conferred to the TPTC by Annex I, which regulates the technical details related to the utilisation and protection of the watercourse. The Committee is entrusted with the review of operating rules of existing dams (Article 4 (6) and 6 (6) of Annex I), the determination of the minimum river flows (Article 5 (2) and 7 (2) of Annex I) and the approval of operating rules for hydropower installations in any of the three states (Ashton et al. 2006).

The TPTC also plays an important role in compliance assurance. A compliance assurance mechanism is a set of rules and procedures aimed at assessing, regulating and ensuring compliance. In the Incomaputo-Agreement elements of compliance control mechanisms are contained in various provisions of the agreement. Article 4, requires the parties to the agreement to "establish comparable monitoring systems, methods and procedures' (Article 4 (h)) and to "promote the implementation of this Agreement according to its objectives and defined principles" (Article 4 (j)). The agreement furthermore requires the TPTC to meet at least twice a year "for the purpose of implementation of this Agreement" (Article 5 (3)).

The agreement places particular emphasis on the TPTC's role in compliance assurance regarding pollution prevention and control. Article 8 (1) (e) requires the TPTC, through committee resolutions, to implement a regular monitoring programme to meet water quality objectives and to review the list of polluting substances in order to enable compliance (Article 12 (3)). These provisions are complemented by Article 12 (6) and (7), which stipulates requirements for the exchange of information relating to the water quantity and quality necessary for implementation.

The substantial delegation of powers to the TPTC makes the institution the key player in the implementation of the Incomaputo-Agreement. The fact that the parties have agreed to give the watercourse institution such wide-ranging powers shows a willingness to put their frequently expressed commitment to regional solutions into practise. As such, the Incomaputo-Agreement is in line both with the legal framework as well as the spirit of the SADC Protocol.

13.5.4 Domestic Enforcement

The legal framework for transboundary water resources management set by the SADC Protocol and basin-wide agreements is complemented by the domestic laws of the member states. Essentially, the effective implementation of international agreements depends on the interaction between international and national laws, as enforcement on the national level has to make use of the instruments of national laws. As said, many countries of the SADC region have reformed their domestic water laws and policies (or are currently in the process of doing so).

One tendency of the new domestic water laws is the devolution of water management to the lowest possible levels, in line with the principles of IWRM. Examples of this are the establishment of Catchment Management Agencies (CMAs) in South Africa, which are envisaged to take over most water resources management responsibilities from the Department of Water Affairs and Forestry (DWAF) over time. Similar examples of management devolution can be found in Namibia, where the Water Resources Management Act (No 24 of 2004) provides for basin management committees, local water user associations and water point user associations, and Zimbabwe where Catchment Councils are largely in charge of water resources management. Botswana is in the process of developing a new water act and the first draft bill (Republic of Botswana 2005) provides for the establishment of water management areas along catchment boundaries.

Taking the example of the Incomaputo-Agreement, South African authorities need to ensure that no more water than the country's agreed share is allocated to users within its territory. If South African domestic law would not contain any or make insufficient provisions that permit authorities to regulate the allocation of water, it would not be possible to prevent water users from using as much water as they like with the possible result being that South African users use more water than allocated to that country as its agreed share. In that case, South Africa would be in breach of an international agreement without there being a violation of South Africa's domestic law. In practise, the challenge of effectively implementing international agreements became evident in the past where, during drought times, South Africa did at times not meet the terms of the agreement (the agreed minimum flow into Mozambique). For this reason the national water law must recognise the country's international obligations and ensure compliance with them. It is here where Section 2 (i) of the National Water Act is of central importance. The provision forms the link between South Africa's international obligations and domestic enforcement. It enables, and obliges, the domestic authorities (DWAF and CMAs) to include the international obligations in the equation when making allocation or licencing decisions domestically. Legally, the framework for the management of water resources on a transboundary scale is only complete where international and domestic law are interlinked as in the above example and together provide the means for the effective implementation and enforcement of international agreements. This aspect should therefore be given greater attention in the various capacity building initiatives aiming at creating improved capacity for transboundary water resources management in the region and beyond.

13.6 Conclusion

The countries of the SADC have developed a far-reaching policy and legal framework for the management of the region's shared water resources. The ongoing strengthening of Shared Watercourse Institutions, such as ORASECOM and others, and the increasing role they play in basin management indicates that the countries of the region are committed to translating the policy and legal framework into cooperative action on the ground. Importantly, in this effort they are guided by the principles of international water law as enshrined in the SADC Protocol (and the UN Convention) and the various basin-wide agreements drafted in accordance with the former. The Incomaputo-Agreement is at present the most comprehensive basin-wide agreement and may provide a good example for other basins that aim at developing similar agreements in the long-run.

Domestically South Africa's commitment to the cooperative management of its shared water resources based on international water law is reflected in the country's national water policy and the National Water Act with its specific requirement to ensure that international obligations are met.

Following its transformation to democracy South Africa has moved away from the previously preferred model of bilateral cooperation (only) and is now fully committed to regional integration and managing its shared water resources on a basin-wide scale based on the modern principles of international water law and policy. More so, South Africa, together with its partners in SADC, is involved in continental activities (under AMCOW and others) aiming at strengthening cooperation over shared water resources in Africa as a whole, thereby recommending the SADC advances as a possible model for an African approach.

References

AMC (2006) Declaration and action plan of the African ministers' conference on Hydropower and sustainable development. African Ministers' Conference. http//www.hydropowerconference2006.co.za. Accessed Retrieved 16 Sept 2006

Ashton PJ, Earle A, Malzbender D, Moloi MBH, Patrick MJ, Turton AR (2006) A Compilation of all the international freshwater agreements entered into by South Africa with other states. WRC Report No. 1515/1/06. Pretoria, Water Research Commission

DWAF (Department of Water Affairs & Forestry) (2004) Introduction to the Orange River Basin. Pretoria, Department of Water Affairs & Forestry. http//www.dwaf.gov.za. Accessed 7 June 2008

Earle A, Malzbender D, Turton A, Manzungu E (2005) A preliminary basin profile of the Orange-Senqu river. African Water Issues Research Unit (AWIRU), University of Pretoria, Pretoria

GWP (2005) Toolbox integrated water resources management – update August 2005. Global Water Partnership, Stockholm

Heyns P (1995, November). Existing and planned development projects on international rivers within the SADC region. In: Proceedings of the conference of SADC ministers responsible for water resources management, Pretoria

Heyns P (2004, January) Achievements of the Orange-Senqu river commission in integrated transboundary water resource management. Paper presented at the General Assembly of the International Network of Basin Organisations in the Martinique

Malzbender D, Earle A (2007) Water resources of the SADC: demands, dependencies and governance responses. Paper presented at the Institute for Global Dialogue's (IGD) and Open Society Initiative for Southern Africa's (OSISA) workshop on "Natural Resource Dependence in Southern Africa: Towards equitable, accountable and sustainable use". Institute for Global Dialogues and Open Society Initiative for Southern Africa, Pretoria

Republic of Botswana (2005) New Water Act. First Draft Bill, Government of Botswana

Republic of Botswana, Kingdom of Lesotho, Republic of Namibia & Republic of South Africa (2000) Agreement between the governments of the republic of Botswana, the kingdom of Lesotho, the republic of Namibia and the republic of South Africa on the establishment of the Orange-Senqu river commission

Republic of Mozambique, Kingdom of Swaziland & Republic of South Africa (2002) Tripartite interim agreement between the republic of Mozambique and the republic of South Africa and the kingdom of Swaziland for co-operation on the protection and sustainable utilisation of the water resources of the Incomati And Maputo watercourses

Republic of South Africa & Kingdom of Lesotho (1986) Treaty on the Lesotho Highlands water project between the government of the republic of South Africa and the government of the Kingdom of Lesotho. Republic of South Africa, Kingdom of Lesotho, Maseru

RSA (1996) The Constitution of the Republic of South Africa (Act No. 108 of 1996). Government Printer, Pretoria

RSA (1998) National Water Act. Government Gazette Vol. 398. 26 August 1998. Office of the President, Cape Town

SADC (2005) Regional water policy. Southern African Development Community, Gaborone

Turton AR (2004) The evolution of water management institutions in select southern African international river basins. In: Tortajada C, Unver O, Biswas AK (eds) Water as a focus for regional development. Oxford University Press, London, pp 251–289

UN (1997) Convention on the law of the non-navigational uses of international watercourses. United Nations General Assembly, New York

Vinogradov S, Wouters P, Jones P (2003) Transforming potential conflict into cooperation potential: the role of international water law. UNESCO/ IHP/WWAP/ IHP-VI/ Technical document in Hydrology/PCCP series No 2

Chapter 14
Lessons and Conclusions

Barbara Schreiner and Rashid Hassan

Abstract This book discusses some issues of water policy and its implementation in South Africa. This chapter highlights some key, overarching issues related to the implementation of water policy in South Africa. These issues are discussed in the context of good water governance, based on three principles: a professional bureaucracy; predictable, open and enlightened policy making; and a strong civil society active in public affairs. The chapter closes with some discussion on some of the key challenges and priority actions in moving forward.

Keywords Water governance • Professional bureaucracy • Civil society • Policy

14.1 Introduction

This book set out to discuss some of the issues of water policy and its implementation in South Africa. While it is not possible for a work of this nature to be comprehensive, a range of the key issues have been dealt with in the preceding chapters. The hope is that these chapters will provide some insights to water practitioners across the world, particularly those in developing countries, and in South Africa, on some of the challenges and lessons arising from the South African water sector. Within South Africa itself, it has been useful to take the time out to reflect on these issues.

It is not the intention of this chapter to repeat the detailed analysis of the previous chapters, but rather to draw out some key, overarching issues with regard to the

B. Schreiner (✉)
Pegasys Strategy and Development, Pretoria, South Africa
e-mail: Barbara@pegasys.co.za

R. Hassan
Centre for Environmental Economics and Policy in Africa, University of Pretoria, South Africa
e-mail: Rashid.hassan@up.ac.za

B. Schreiner and R. Hassan (eds.), *Transforming Water Management in South Africa: Designing and Implementing a New Policy Framework*, Global Issues in Water Policy 2,
DOI 10.1007/978-90-481-9367-7_14,

implementation of water policy in South Africa. These challenges arise from a context that is policy rich, but relatively poor in implementation capacity.

The issues highlighted in this chapter are set within an analytical framework of good water governance. Good water governance is based on three key factors: predictable, open and enlightened policy making, a professional bureaucracy acting in the public interest, and a strong civil society active in public affairs (WRC 2005). It is within this framework that implementation of water policy in South Africa is discussed.

14.1.1 Predictable, Open and Enlightened Policy Making

It is evident from the preceding chapters that South Africa has excellent policies and legislation in place at the macro level, in relation to both water services and water resources management. The policy and legislation were developed through open and participatory processes, new to the water sector in South Africa. International expertise was also called on to ensure that the policy reflected international best practice at the time.

What has, equally, been made clear in the previous chapters is that the key challenge lies not in the policy so much as in the translation of this policy into practice. Much has been achieved in the implementation of the policy, yet, as highlighted in many of the chapters, there are major challenges in implementation arising, inter alia, from lack of capacity, over-ambitious and highly technical interpretation of policy, and the desire to do too much at one time.

The challenge of implementing a complex suite of water policies has been compounded by the demands of national policies outside the water sector. The water sector does not operate in isolation from broader forces in the society, and policies such as those requiring affirmative action in appointments and black economic empowerment (BEE) in appointments of consultants and contractors, added another layer of challenges to the already complex work of the water sector. Affirmative action and BEE are critical and necessary policies in the South African context, and they brought to the water sector a new perspective and drive for equitable water allocation and service delivery. Equally, however, they also resulted in loss of experience and institutional memory at a time when the water sector was called on to implement new and complex policies. The racially biased educational policies of the apartheid state also contributed to capacity challenges through the limited number of technical graduates within the black population – skills necessary for an effective water sector. The massive spectrum of policy changes happening simultaneously in South Africa has added to the challenges facing the water sector.

The water sector in South Africa is also governed by two separate White Papers and associated legislation – one on water services and one on water resources. While there were valid reasons for this approach at the time of the development of the policy, the approach taken within DWAF resulted in two very distinct streams within the water sector, one dealing with water services and one dealing with water resources.

A lack of interface between the two streams resulted in gaps, overlaps, and lack of alignment in policies and programmes. It also resulted in a lack of understanding between the two streams of each other's work. The development of a strong interface between these two policy arenas is only now happening. This will, one hopes, overcome some of the policy and implementation gaps previously experienced, such as the lack of a combined and comprehensive approach to the managing waste water treatment works from both a water resources and water services perspective.

The internal policy and implementation challenges were exacerbated by poor alignment of policies between departments, such as in water services delivery (Departments of Water Affairs and Forestry, Housing, Provincial and Local Government and National Treasury), water allocation reform (Departments of Water Affairs and Forestry, Agriculture and Land Affairs), and water resources protection (Departments of Water Affairs and Forestry and Environmental Affairs and Tourism). Improved alignment would have enabled more effective use of resources and implementation.

14.1.2 Professional Bureaucracy

The period since 1994 has seen significant changes in the water bureaucracy in South Africa. Such changes include the redefinition of local government boundaries, the incorporation of former homeland water administrations and "white" South African administrations into the Department of Water Affairs and Forestry, and the establishment of CMAs. At the same time, the affirmative action policy saw significant shifts in the staffing of these institutions to better represent the racial and gender demographics of the country.

The Department of Water Affairs has been transformed from a primarily technical department to one in which social and environmental issues are of equal importance. However, the slow rate of transformation of WUAs and establishment of CMAs linked to the drain of skills and institutional memory from the civil service has resulted in severe capacity constraints in the implementation of the extremely progressive policy and legislation. A similar drain in capacity has taken place at the municipal level, affecting the ability of municipalities to deliver, operate and maintain effective and sustainable water supply and sanitation.

Thus, while demands on the bureaucracy have increased significantly because of the new policy and the need to deliver water and sanitation to millions of unserved South Africans, actual institutional capacity has been shrinking. A number of programmes have been put in place to turn this around, such as the establishment of a Learning Academy within the Department of Water Affairs and Forestry, the Siyenza Manje programme run by the DBSA to provide technical support to municipalities, and a range of capacity building and training programmes. However, it has become increasingly clear that, alongside continuing attempts to build capacity, it is necessary to tailor the programmes and targets according to the capacity to deliver. Less ambition may result in better delivery.

14.1.3 Informed Civil Society

The period since 1994 has seen both positive and negative developments in relation to informed civil society. A relatively strong NGO sector was weakened by the loss of donor support and staff after 1994, as international donors transferred their support to the new, democratic government. At the same time, new processes were put in place to open up policy and planning processes to stakeholders – as mandated by various pieces of legislation, including the National Water Act.

Wide consultation on the National Water Resources Strategy, for example, brought water resources management into the domain of a range of stakeholders for the first time, a key step in democratising water management and an equally key step in developing an informed and effective strategy – public input significantly altered the content of the NWRS.

Public awareness campaigns, public consultation programmes, and the role of organisations such as scientific and research organisations, NGOs and private sector bodies, have raised involvement in and understanding of water issues. The media have played a critical role in this regard as well. However, at the local government level, lack of responsiveness by elected officials to the needs and aspirations of their communities have contributed to protests against poor service delivery. It is argued in this book that a closer relationship of accountability is needed between local government representatives and their communities in order to ensure the sustainability of service provision.

A particular challenge in South Africa has been ensuring that women and black communities are well represented in participatory processes in the water sector. To support this DWAF developed a set of guidelines for participation in water processes. However, because of the huge and continuing inequities in South African society, ongoing commitment to ensuring the participation of the vulnerable and marginalised remains critical.

14.2 Issues in Moving Forward

The very real threat of global climate change, combined with the increased impacts on water resulting from demographic and economic changes, brings increased challenges for the South African water sector. Already, South Africa is struggling to implement successfully its very progressive policy and legislation.

To meet its objectives, it will be necessary for the water sector in South Africa to focus on three key steps:

- The identification of priority programmes: the water sector must determine those programmes that are critical and non-negotiable for the achievement of better water management.
- In tandem, the water sector must determine what programmes currently being implemented can be halted or postponed so that limited capacity can be directed to the priority programmes instead.

– Thirdly, the water sector must determine how to simplify and streamline those priority programmes that are overly resource intensive, detailed or complex, so as to enable effective implementation with reduced resources.

This approach will enable the alignment of programmes to available capacity, not the other way around. The section below discusses these three factors: priority actions, programmes requiring simplification, and the delaying of some programmes.

14.2.1 Priority Actions and Simplification of Processes

Effective water resources and water services planning, at national and local levels, and the integration of water planning with social and economic planning remain the foundation of effective water resources management. Such planning is necessary to understand and implement the activities required to reconcile demand and supply. The actual integration and alignment of planning processes remains a challenge, with the interface between resource and services planning a particular challenge. DWAF has responded to this latter challenge with a programme to prepare water resource plans for all municipalities over the next few years. However, the issue of planning capacity at local and national level remains a concern.

Water conservation and demand management is an increasingly important element of the reconciliation of supply and demand and increased resources, both human and financial, will need to be allocated to a national water conservation and demand management programme. Such a programme should include improved maintenance, operation and rehabilitation of water infrastructure, both bulk and retail, as well as communication and awareness raising, and improved enforcement of licence conditions and standards. The development of new infrastructure, both water resources and water services, must also continue, particularly in the light of the need to provide universal water services coverage and the water resources challenges of climate change.

A key part of the challenge of effective management and rehabilitation of infrastructure relates to the management of return flows from municipal waste water treatment works. At the same time, pollution from mines, industry and agriculture also needs to be managed more effectively. It is critical that a strong programme of improved compliance monitoring and enforcement is put in place and that illegal water use and effluent discharge are brought under strict control. Without strong compliance and enforcement a range of objectives in the water sector such as protection of aquatic ecosystems, reallocation of water and enhanced water use efficiency will not be achieved.

In order to achieve these key programmes, it is important to have stable institutions. South Africa has been in the process of institutional restructuring in the water sector since 1994. In the water sector this has seen internal restructuring of DWAF, the (slow) establishment of CMAs, the transformation of Irrigation Boards into Water User Associations and the establishment of new WUAs, and the transformation of local government. Such change, while important, is also destabilising and demoralising when it continues for too long. The efficient finalisation of restructuring would provide a stable platform from which key programmes of the water sector could be implemented with improved concentration and effectiveness.

The Water Allocation Reform programme remains critical to meeting the transformation objectives of the South African government, but, as set out in Chapter 4 it is not without its challenges in implementation. The programme, as currently devised, is highly resource intensive and technical, and as a result, is both slow and open to legal challenge on a range of issues. Alternative approaches and options that can be applied either instead of or as well as the current approach may be necessary to facilitate faster and more efficient achievement of the objectives of the programme. The reallocation of water, along with the reallocation of land, is an important part of the redress of historical injustices in South Africa.

To support both the WAR programme, and the authorisation of water use, more efficient licencing procedures and better use of General Authorisations are required. While much work has already been done by DWAF in streamlining licencing procedures, the process is still slow and cumbersome. The DWAF Annual Report of 2007 records a backlog of 1,300 licences, with only 90 having been issued during the year.

One of the reasons in the delay of licences has been cited as the slow pace of reserve determinations. Once again, while this has changed considerably over the past 10 years, too much emphasis is being given to the determination of reserves and too little to their implementation. Quicker determination of reserves with a stronger focus implementation would result in more impact than the continued focus on attempting to determine accurate and precise reserve requirements.

Such focus on priority areas, and on implementation rather than policy and strategy, will enable the South African water sector to deliver the promise of its remarkable water policy and legislation to the people on the ground. Such delivery will be the true measure of the success of the policy reforms.

Reference

WRC (2005) Strategic review of current and emerging governance systems related to water in the environment in South Africa. Pegasys Strategy and Development. Water Research Commission, Pretoria, July 2005

Subject Index